Machinic Ontology

Mehdi Parsa

Machinic Ontology

palgrave
macmillan

Mehdi Parsa
Bonn, Germany

ISBN 978-3-031-99560-6 ISBN 978-3-031-99561-3 (eBook)
https://doi.org/10.1007/978-3-031-99561-3

This Palgrave Macmillan imprint is published by the registered company Springer Nature Switzerland AG.
The registered company address is: Gewerbestrasse 11, 6330 Cham, Switzerland

Preface

A book is like a brain, a neuronal network composed of countless numbers of cells that cluster together to form circuits, columns, and various interconnected functional areas. Or, it is like a cosmic web of the universe composed of clusters of galaxies, filaments, matter sheets, and voids. In any case, it has no center and no inside, and its title is given to it by a stepfather (though it doesn't make any difference if it is a real father). "A book itself is a little machine" (ATP, 4). This book also has no center and no principal idea, but rather is a network of problems that form connectomes and synapses. I'm not its father, but its child. The authors who try to structure their book very well are the ones who believe in the essentiality of names and the transcendence of forms. They try to hide that ideas are the children of coincidences, and problems are just other names for events. They write in a manner to pretend that the sequences and connectomes are necessary. But let us at least diminish this pretention. This has nothing to do with nihilism, relativism, or post-modernism. It does not mean that everything is everything. It has to do with naturalism. A brain is formed by coincidences. Nonetheless, it provides a complex set of logical functions, although it always remains open to new coincidences. The connections are evental but objectively real.

Thinking is a direct consequence of the networked nature of the brain. In a loose, albeit not apathetic, sense of the term, it can be asserted that any network has the potential to generate outcomes that possess the fundamental characteristics of thought. Wherever there is time, there is thought. And time exists only where space can be bent, shaped, and structured. A philosophical tome can be regarded as a contribution to the ongoing development of a historical network, exclusively reliant on collective knowledge. History encompasses far more objective knowledge than any individual philosopher. When viewed through this lens, a whole brain becomes akin to a synaptic connection, and a book becomes nothing but a mere chapter.

This book is an effort to display the becoming-concept of the machine-metaphor. Its aim is to explore the intersections between this becoming and the development of the concepts of life, organism, and technics. It seeks to introduce universal machinism as a metaphysical foundation with specific ethical and political implications: Machinism without life is tantamount to fascism, and life devoid of machinism verges on fanaticism. A machinic ontology proposes that there is no stable inside, the body has no head, a society needs no leader, and the brain has no center. While undoubtedly a Deleuzoguattarian idea, we endeavor to explore its origins and echoes across various thinkers and domains. In this context, we examine and analyze the concepts of monstrous machines, natural machines, irrational machines, and living machines as they appear in the works of diverse thinkers across various fields.

The first Appendix is already published in Mehdi Parsa, "Max Ernst and the Schizoanalysis of Nature", *Epoché Magazine*, Issue 74 August 2024.

Bonn, Germany

Mehdi Parsa

Contents

Abbreviations of In-Text References

AO	Deleuze, Gilles and Guattari, Félix. *Anti-Oedipus: Capitalism and Schizophrenia*. Trans. Robert Hurley, Mark Seem and Helen R. Lane. Minneapolis: University of Minnesota Press, 1983.
ATP	Deleuze, Gilles and Guattari, Félix. *A Thousand Plateaus: Capitalism and Schizophrenia*. Trans. Brian Massumi. London and New York: Continuum, 2004.
Cinema II	Deleuze, Gilles. *Cinema II: The Time-Image*. Trans. Hugh Tomlinson and Robert Galeta. Minneapolis: University of Minnesota Press, 1997.
Communist Manifesto	Marx, Karl and Engels, Fredrick. *Manifesto of the Communist Party*. Peking: Foreign Languages Press, 1970.
Courage	Foucault, Michel. *The Courage of the Truth (The Government of Self and Others II) 1983–1984*. Trans. Graham Burchell. London: Palgrave Macmillan, 2011.
DR	Deleuze, Gilles. *Difference and Repetition*. Trans. Paul Patton. London and New York: Continuum, 2004.

Individuation	Simondon, Gilbert. *Individuation in Light of Notions of Form and Information*. Translated by Taylor Adkins. The University of Minnesota Press, 2020.
LoS	Deleuze, Gilles. *The Logic of Sense*. Trans. Mark Lester. London: The Athlone Press, 1990.
Mode	Simondon, Gilbert. *On the Mode of Existence of Technical Objects*. Translated by Cecile Malaspina and John Rogove. Univocal Publishing, 2017.

1

Introduction

There has been a growing trend in various fields of contemporary philosophy, sociology, psychology, linguistics, and other areas of the human sciences that involves replacing classical concepts such as the object, the individual, the mind, the signifier, and others, with that of the machine or other related concepts, such as the network or the artificial brain. This shift in perspective is a response to the significant changes in our world caused by scientific and technical advancements. These changes have not only impacted our physical surroundings but also our worldviews and philosophies. From seventeenth century onward, we live in a world that is increasingly shaped by machines. Unsurprisingly, this change was met with nostalgia, negative reaction and resistance, and then became an inevitable reality that everyone must accept (our bloody realism). This shift is reflected in the way we used to approach both the natural world and human societies. As Gerald Raunig observes, "In this era [17th century] there was a proliferation of metaphors of man as machine, of the state as machine, of the world as machine" (Raunig 2010, 20). Then comes philosophy to the game, where machine is viewed as a metaphor or concept to describe the essence of reality, given the broad range of possibilities opened up by modern scientific developments. The evolution of the natural sciences in the nineteenth and twentieth centuries has incited

M. Parsa, *Machinic Ontology*, https://doi.org/10.1007/978-3-031-99561-3_1

modifications in the machine concept. Machines have evolved from being solely man-made technical instruments to now being seen as a philosophical tool, providing a fresh perspective on how we view the world: from the Laplacian mechanical machine-cosmos to the malleable spacetime machine of General Relativity. The concept of the machine is no longer limited to technical machines, but has expanded to include not only physical probability but also life itself. Machines have emerged to assume the role that the concept of organism used to play. Anything that was once described as an organism is now referred to as a machine. Now, machines not only surround us but have become infectious monsters that transform everything, including nature, into machines. As a result, the machine has become the prevailing concept of our time and requires a new ontology. In the realm of the human sciences, we no longer live in the era of structures or organisms but rather in the era of machines. Structural or social constructivism is being replaced by machinic constructionism.

What is philosophy in the age of machines, machinic intelligence, and machinic thinking? Philosophy must create a concept or a collection of concepts from this new reality to model it in a way that is distinct from the conventional logic of substances and properties. It's not just that we are surrounded by an abundance of machines that shape our environment, but modern sciences are transforming everything into machinic entities such as networks, orbitals, clouds, and fields. Philosophy would fall behind the sciences if it continues to limit itself to discussing the ontology of things or simple objects on the one hand and living beings and subjects on the other. How can philosophy address not only the consensual world of ordinary objects and problems but also the amazing worlds that modern physics, chemical biology, ecology, cosmology, and neurology have revealed to us? How can philosophy approach these worlds, or ontological fields, without reducing them into a mere totality (*Being*) alien to each of them? How can we see things otherwise, though still philosophically, in terms of the concept of machine and the variety of other concepts and expressions related to it, so that the localities and singularities are not reduced into generalities and universalities? How can we liberate ontology from the concept of being? How can the *ipseity* of things be liberated from the categories of object and subject? How can a philosophical concept of machine assist us in achieving this goal? Furthermore, how can philosophy utilize this invention to establish a connection between the scientific expansion of the world and our comprehension of ethical-political life? Can Hobbes' concept of society as an organic body be reconciled with Descartes' notion of nature as a machine? How is it possible to have a machinic social ontology? How seeing otherwise leads to living otherwise? How might the idea of machinic life connect the notion of non-organic life of nature with the concept of social life? To address these issues, it's necessary to discuss the nature of modern science and its relationship with society and philosophy.

The history of technicity along with the Hegelian facticity of human life and knowledge leads philosophy to recognize agency and productivity in the non-human world. There have been numerous attempts in recent decades to articulate this ontological productivity, ranging from various iterations of new materialism to the so-called new realisms, posthumanism, and other related movements. By studying the history of science and technology, we can observe that nature possesses a particular kind of creativity that is distinct from the logocentric creativity of human mind and its panpsychist projection onto the vibrant matter. Building upon Hegelian thought, which posits that if there is spirit within human beings and humans exist within nature, then spirit must exist throughout nature, one can reformulate this idea to suggest that if technicity is a part of human history and human history is a part of natural history, then natural history can be understood in terms of an extended notion of technicity. In short, nature is creative in the machinic sense of the word. Then we can return to humans and conclude that even the human thinking needs to be reformulated in terms of this new machinic creativity: the brain as a natural artifact, an artificial brain fabricated by nature.

Consequently, we arrive at the point of considering human societies as operating differently from what is traditionally understood as being composed of conscious human agents. While there is a consensus on the fact that non-human reality is endowed with agency, the nature of this agency remained obscure. Some proponents argue for the irreducibility of matter to any form of creativity, which they believe to be a uniquely human quality. They instead advocate for the notion of absolute logical contingency as the principal ontological characteristic of matter (Meillassoux

2008). Conversely, others attribute vitality and liveliness to matter and defend a type of vital materialism (Bennett 2010). Our claim here is that the productivity of nature is machinic, and this machinism can serve to diverge from both vitalism and mechanism. It is one of our major objectives in this book to demonstrate that natural machines are neither organic nor mechanistic. Natural machines possess a primordial technicity that is more fundamental than modern human technics and its calculability. They also possess a non-organic life that cannot be derived from living beings but can be understood in terms of what constitutes the exterior surface of organisms, that which connects them to nature in its purest form, transgresses their supposed purposiveness, and is referred to as the death drive or *Thanatos* in psychoanalytic terminology. Machines are *thanatic* organisms. The machinic life contains no living substance and no organic rigidity but implies a technical automaticity which is essential to any networked actor.

This takes us to the societal aspect of the present study. This account of machinic ontology cannot be pursued without invoking an *a priori*, or pre-individual, sociality that dismantles substantial elements and transcends individualism. As a result of Hegel's social revolution in philosophy, it is widely believed that the social part of human existence should be considered *a priori* in relation to our existence as individuals. It is incorrect to start with the characteristics and dimensions that belong to our individuality in order to make sense of our deepest issues and concerns. In Nineteenth and Twentieth century European philosophy, this belief appeared in a variety of ways, from Marxist ideas of commonality to the emphasis on the function of language, structure, the social unconscious, the ontologies of power, and the philosophies of life. The initiatives to prioritize the social over the individual can be roughly categorized into abstract and concrete levels. The abstract side includes the formalist, structural and relational features of that priority. The concrete side rather focuses on the pre-individual conceptions of life and power. What seems demanding in this situation is introducing a revision of the concept of machine that can bridge these two sides, focusing on the priority of the social over the individual without falling into structuralism or vitalism. Consequently, human societies as machines that are composed of flows and their articulations operate within the same metabolism of natural machines. Social agency is more similar to natural agency—the agency of non-organic life—than to conscious individual agency or representational state agency.

Thus, we will employ the notion of the machine as a specific concretization and deconstruction of the concepts of structure and form. This is made possible by the problem of life, not as a substance inherent in living beings, but as a problem that radically transformed our scientific perspective. We learned already from Nietzsche how spiritless our theories, structures and formulations are. We learned that we must search beneath the structures their genealogies, which is to say, how living forces can interact and synthesize to produce a system of fluxes, which in turn creates an abstract structure. In other words, how life generates forms. Nietzsche demonstrated how this movement of abstraction has been in motion since the earliest emergence of the living forces, and the so-called forces, far from being individualized, enjoy a certain articulation and sociality in terms of potencies and intensities. Structuralists attempt to argue that the structure precedes the individuals, but they overlook the fact that every structure emerges through a process of abstraction. There exists therefore always a gap in their work between the abstract and ideal structure and the concrete fluxes of life. They fail to reach the notion of a natural agency which is not derived from the conscious human agency. It is therefore a demanding proposal that in order to maintain a real connection between the abstract form and the concrete force the notion of structure must be replaced with the notion of machine as the material ground of structures. Machines are power-full, or motorized, structures. In this condition, one can provide an explanation of sociality which is less structural and more material and forceful; it is indeed based on the idea of society as a living machine. Therefore, the present study centers on the relationships between subjectivity, consciousness, the unconscious, life, and machines. It attempts to formulate and respond to the question: Is subjectivity necessarily dependent on living organisms, considering that structuralism and post-structuralism isolate subjectivity from living organisms, relating it instead to signifiers, traces, memories, and information? Is a machinic subjectivity possible, considering the impossibility of intimacy of consciousness, reducing the so-called autobiographical self into mere biographical selves?

In the current work, we make references to a few philosophical texts that each in their own way study the relationship between the concepts of life, nature and machine and try to extrapolate relevant political implications from this relationship. The majority of texts that we examine here offer a critique of the classical idea of machine as manmade, monofunctional and mechanical, as well as a critique of the classical idea of life as organized and containing the vital substance. They try to redefine the concepts of machine and life on the basis of the historicity of science and technique and the fact that the history of technicity and scientificity takes place within the natural history. They, therefore, attempt to provide a description of machine that goes beyond manmade tools or mechanical calculability and is compatible with an understanding of network which is an essential component of living organisms, physicochemical systems, and cosmic structures. On this basis, we will introduce the thought of machinic ontology which takes the concept of machine as a useful philosophical instrument to describe the nature of existence in its diverse manifestations including both dead and living bodies, natural systems and societies. This can provide us with an alternative for the ontology of things and substances which regards the world as a sum of mechanical objects and organic living entities. Here is the claim that we will try to defend: We used to argue that if nature has the capacity to generate the spiritual human beings, it must be endowed with a kind of spirituality. The new argument is that if nature is capable to produce machines, it deserves to be called machinic. But as in Hegel the absolute spirit is different from the human spirit, the ontological machine too would be different from the manmade machine. This is the new machinic materialism which is based on technics and the non-spiritual life. Machinality is the characteristic of the productive force of nature, or rather, the life of nature. Nature is living is a special sense; its life is machinic.

To accomplish this aim, we derive from the texts that we study the notions that serve us to formulate this machinic ontology: monstrous machine (George Canguilhem), transindividual machine (Gilbert Simondon), plastic machine (Catherine Malabou), abstract machine and war machine (Gilles Deleuze and Félix Guattari), and bio-political machine (Michel Foucault and Giorgio Agamben). Each of these conceptions would enable us to further develop certain aspects of the machinic ontology. Respectively: the fundamental dependency between life and machine in Canguilhem, its extension to the ontogenetic level in Simondon, expanding on its (deep) historical aspect and connecting it to neurotechnicity in Malabou, describing the logic of its functioning as the logic of multiplicities and revealing its revolutionary nature in Deleuze and Guattari, and finally analyzing the ethical-political consequences of this machinic ontology in Foucault and Agamben.

This book will consist of seven chapters and two appendices that have a kind of kinship affinity with the chapters. After this introductory chapter, the second and third chapters will focus on George Canguilhem and Gilbert Simondon, two prominent figures in twentieth-century French philosophy of science, who explored the relationship between the concepts of life and machine. Building on Descartes' *Treatise on Man* (*L'Homme*, 1662), Canguilhem introduces the concept of living machines. By integrating this idea with his view of life as lacking inherent vitality, as pathological and monstrous, we derive the notion of the "monstrous machine" to describe the extraordinary products of nature. This is based on Canguilhem's application of the concept of indeterminacy, derived from the life sciences, to the physico-chemical sciences, leading to conclusions about the notion of natural undetermined or monstrous machines. He proposes the convertibility of life and technics, exploring the essence of technology within the ethnological development of man as a component of life's technical evolution. Therefore, Canguilhem's deconstruction of the organism in "Machine and Organism" results in a machinization of the organism, based on its multifunctionality in response to the diversity of milieu-systems and the differentiation of value-systems. As discussed in the fourth section of Chap. 2, although Canguilhem attempts to understand machines in terms of organisms, the organisms have already been machinized—that is, freed from the concept of vital substance. This culminates in a universal view of science that provides an ontological framework.

Gilbert Simondon takes a more overt move in the direction of ontology. In our third chapter, we aim to establish a connection between the concept of the machine in *The Mode of Existence of Technical Objects* and the notion of the transindividual in *Individuation in Light of Notions of Form and Information*. Linking the machine to the transindividual, as the culmination of physico-biological and psychosocial individuation, provides a comprehensive ontogenesis—what can be termed a machinic ontogenesis. This encompasses various levels of machines—physical, biological, psychosocial, and technical—where conventional technical machines represent only one of its superficial aspects. Simondon defines machinality in terms of relationality, embedded in the very fabric of being. Life can generate technology because the entire process of ontogenesis is grounded in this relationality, which infuses the physical, biological, and other domains of individuation with a machinic essence.

In our fourth chapter on Catherine Malabou, we discuss three of her studies, first a series of lectures that she delivered on the topic of Anthropocene, then her book *What Should We Do with Our Brain?* which is about the plasticity of the brain, and finally her studies on machinic intelligence and AI in *Morphing Intelligence*. We get from her discussions on the Anthropocene the idea of human being as a geological force alongside other geological forces and the fundamental link between the history that we write and the (deep) history that produces us. When we think of human being as a geological force, we are led to believe that the nonhuman geological forces, which ultimately include the earth itself, are not merely mechanical forces, but can be thought of as historical agents. We will try to tie this idea with the plasticity of the brain, a thought that Malabou discusses in *What Should We Do with Our Brain?* Malabou restricts herself to discussing the plasticity of the human brain, but we will try to expand on her idea to include the geological context she discusses in her lectures on the Anthropocene. The plasticity of the brain, which underlies its non-centrality and liberating it from the commanding role, also contributes to the dehumanization of the brain and its expansion as the brain of deep or geological history. This gives us a critical step for thinking about a same machinality in the operation of both the brain and nature. With regard to machine intelligence, Malabou's comments in *Morphing Intelligence* will provide us a guiding idea that shifts from the human brain to the machine brain and its plasticity, which makes it being opposed to mechanism, which is traditionally thought of as the element of machinality.

The most direct and comprehensive explanation of machinic ontology can be found in the two volumes of *Capitalism and Schizophrenia* by

Deleuze and Guattari. In these volumes, they boldly assert that everything is a machine, and that every machine can be transformed into a war machine. Our fifth chapter primarily examines the concepts of the abstract machine and the war machine in relation to the intersection of desiring-production and social production. In the first volume of these works, *Anti-Oedipus*, Deleuze and Guattari characterize human bodies and societies as schizophrenic machines composed of flows of unconscious desire. The interchangeability of desiring-production and social production forms the foundation of a machinic ontology, where both the living body and society operate according to the same logic. This logic can be referred to as the logic of multiplicities. In these volumes, machines are defined beyond the dualism of mechanism and vitalism, endowed with a kind of non-vitalist molecular life that enables them to operate at any level or scale of reality. Deleuze and Guattari argue that the abstract machine stands in contrast to both subjectivities and their manufactured mechanical machines, as well as to the orthodox scientific interpretations of these artifacts, which are projected onto the realm of societies and nature. Consequently, every abstract machine is also a war machine that engages in theoretical and practical battles against the State apparatus and its institutionalized science. This gives the machinic ontology a profound political and revolutionary function.

In our sixth chapter, we examine Foucault's concept of "true life" in relation to his idea of biopolitics, ultimately assessing how machinic ontology might reshape our ethical and political existence. Our primary objective is to transition from Canguilhem's concept of the machinization of life to the idea of biopolitics, in which the social machine operates through modes of life and aesthetics of existence. Foucauldian biopolitics, as will be demonstrated, would be inconceivable without Canguilhem's reexamination of biology through the lenses of machinality and monstrosity. Thus, our examination of Foucault will address the ethical issue of self-governance as the foundation for politics rooted in dominance and obedience, aiming to replace it with an ethics-politics grounded in social resistance. To achieve this, we will focus on the series of lectures Foucault delivered at the Collège de France in 1983–1984 (*The Courage of Truth*). We will also examine Malabou's interpretation of the late Foucault as an anarchist, placing it in dialogue with Foucault's critique of democracy. In the final section of the chapter, we will concentrate on Foucault's concept of *dispositif* and Giorgio Agamben's interpretation of it in his work "What is Apparatus?" We conclude that Foucault's notion of dispositive, and Agamben's reading of it, indicate an attempt to search for an account of sociality which is not structural and representational. As will be seen, this implies the replacement of machinic affect for structural effect.

Finally, after reviewing the relevant references, we will conclude that machinic ontology can give philosophy a powerful instrument for returning to its ontological endeavor in the wake of criticisms of the classical ontology of essences and substances, the complete rejection of ontology following Kant's epistemological turn, and sporadic attempts to construct anthropomorphic ontologies based on subjectivity. Nevertheless, the machinic ontology does not imply rejecting epistemology. It is not a returning back to pre-critical ontology or a flat ontology of simple objects. This is why it is important to begin with Canguilhem and to insist on the concept of bio-machines, through reconfiguring the problem of the relationship between knowledge and life, and to close our study with Foucault's and Agamben's bio-political machines, which extend the problems regarding knowledge and power. There are machines at work everywhere, from living bodies to planetary systems, but this does not mean a reductionist unification. On the other hand, life cannot and should not be defined in terms of living substances. There is no dualism of living and nonliving substances. Both organisms and machines should be describable with a unique logic of multiplicities. Both physicochemical and living systems are composed of micromachines which are not reducible into things or organisms. This is why, in the macro-level too, we insist on keeping the notion of machine for both living and non-living systems. Libido is the physicochemical element of living systems and this would liberate these systems from vitalism as affirming a vital substance to such systems. In other words, in the micro-level, the difference between the living and nonliving collapses and everything turns to be micromachines, and this holds for all systems in all different scales.

Thus, what we aim to advocate here can be deemed a metaphysical ontology. It is metaphysical in the sense that it posits a proposition about the entirety of existence, rather than a specific realm of reality. We believe that it is inherent in the nature of philosophy to make claims about everything. Even the claim that "we shouldn't make claims about everything because we don't have access to everything" contains a claim about everything. What is presented in this book is an ontology in the sense that it posits a proposition about the nature of reality, which is not limited to "our accessibility" or "accessibility in general", although it does not disregard the issue of accessibility. We will approach the problem of knowledge but as something that exists in nature. It is a post-Kantian, or post-critical, ontology which is not discontent with the philosophies that devote themselves with dealing with the conditions of human knowledge. But it is, at the same time, a metaphysical ontology that elaborates on the claim that "everything is a machine". This would be an effort to philosophize, which is to say making claims about everything, without relying on living or non-living substances.

There was some hesitation regarding the placement of the chapter on Malabou when structuring the chapters. Although the thinkers are generally discussed in historical order, there exists a parallel onto-logical order that begins with the machine-metaphor in philosophical descriptions of "everything" and culminates in an examination of the ethico-political implications of this ontology. In this context, we decided to position the chapter on Malabou before the chapter on Deleuze and Guattari. Throughout the chapters, it will become evident that Malabou's concept of plasticity is radicalized through the discussion of abstract machines, ultimately blurring the distinction between plasticity and machinality. In this regard, the concept of plastic deep history sets the stage for discussing Deleuze and Guattari's abstract machines and war machines as ur-machines. Placing the chapter on Malabou before the chapter on Deleuze and Guattari allows us to connect the decentralized concept of the brain—leading to the problematization of decision-making and agency at the personal level—to the ethical-political challenges of social decision-making and political agency, conceptualized as the political body as a plastic brain.

Machines, Organisms, Structures

Two texts are pivotal for our research: Georges Canguilhem's "Machine and Organism" and Félix Guattari's "Machine and Structure". These texts guide us toward the concept of a machine from two opposite directions: a machine is neither an organism nor a structure. In other words, it is defined by the subtraction of organism and structure.

A structure can be understood as an instrument for applying and distributing power at a collective level. Power, in turn, is made up of the vital force or the principle of life which is structured. The primary question of any political philosophy is: what structures life and how can life resist against it? It is a question regarding the nature of power. Life in itself is wild and monstrous, and there is always a controlling force to articulate it, regulate it, and structure it. This entanglement defines the inherent relationship between the notion of machine and the notion of life. A machine, whether natural or social (but never mechanical), is a structure that encapsulates the resisting living-force within itself. In other words, a machine is composed of life in its bonds and connections, and society is a machine constituted by resistance in its bonds and connections. In Deleuze and Guattari's terms, deterritorialization always precedes territorialization. What, then, is the nature of the power of the social machine, or the ethical machine, and so on? At first glance, this may seem like a question of structural or mechanical enslavement. However, our argument is that it is equally a question of emancipation. A compelling example of this can be observed within the primitive and wild inclinations of political movements during the twentieth century, countering the mechanization of societal power. Gerald Raunig, in his work *A Thousand Machines*, along with discussing these kinds of political movements, draws also a connection to the film *Themroc* (1973), where an effort is made to extend the sphere of political revolution to the ethical-oedipal-linguistic realm. *Themroc* starkly portrays the fundamental division between the mechanistic essence of the power structure and the wild, primitive nature of the resisting vital force. Raunig astutely elucidates how at the beginning of the film the kitchen clock, characterized as both a technical and social apparatus, shapes the Fordist social fabric even before the working day commences (Raunig 2010, 10). A clock operates as a technical machine, effectively organizing time as a natural phenomenon. Simultaneously, it plays its role within the social machine by structuring social life as a cultural phenomenon (in the film, it is always through the mother that the clock functions). By seamlessly intertwining these two worlds, the clock creates *one* mega-machine that blends the natural and cultural dimensions. Despite its technical function of turning all subjects into working cogs within the Fordist social machine and projecting the same mechanism onto time and nature, this structuring is far from operating as a smooth metabolism. Resistance, which is as such "primal and productive", implies that the process of structuring is entangled with the application of power.

In his concise text titled "Machine and Structure", featured in *Molecular Revolution*, Félix Guattari elucidates the contrast between the concepts of machine and structure on the basis of the presence and absence of subjectivity. Unlike structures, machines are devoid of a subject or an agent of action due to their automatic nature. Unlike a structure, a machine "remains essentially remote from the agent of action. The subject is always somewhere else" (Guattari 1984, 112). This is a theme we explore throughout our study by introducing machinic or decentered agency, which refers to the non-deterministic nature of living machines. On a political level, it concerns non-subjective or pre-individual processes of decision-making that are inherently societal in nature. Social decision-making is possible only when each individual locates its subjectivity "somewhere else", outside itself. Consequently, within the mega-machine, the constituent machines are not structurally connected. In the absence of an ideology, each machine influences, infiltrates, and even dismantles the other, forming machinic assemblages. Conversely, structures can establish harmonious connections, maintaining a secure spatial-symbolic separation from one another, thereby shaping the image of subjectivity. In the case of the infra-machines, the only possible separation is temporal.[1]

[1] For a discussion of the entanglement of time and space in Einstein's theory of relativity in dialogue with Bergson, Deleuze's contribution to this discourse, and a brief account of Deleuze and Guattari's metaphysics of space-time, see my article, "A Reconstruction of Gilles Deleuze's Contribution to the Einstein-Bergson Debate". *Transversal: International Journal for the Historiography of Science*, no. 17 (December 2024): https://doi.org/10.24117/2526-2270.2024.i17.08

Therefore, formalism, in the sense of separating a machine from its life-engine and transforming it into a structure, deals with an ideological or logocentric abstraction. In the political discourse, representation indicates the formalist separation of the structuring power and the subjugated social life. In other words, a subject or agent of action ought to embody the social structure, thereby enabling the formation of totalities and State apparatuses. Political machinism involves the abstraction of power in the absence of such agency. In this context, we propose a comparison between the concepts of machinic abstraction and logocentric abstraction, introducing the former as our unique philosophical instrument. We will see that, in this sense, abstraction becomes synonymous with resistance, transforming any logic into a "guerrilla logic" (ATP, 19). We then explore the social machine using non-representational onto-logic and non-representational ethics-politics. Our objective is to redirect the focus of classical Marxist discourse from the interaction between machines and individuals towards the interplay between the natural machinery and life itself. In line with this reorientation, the machine is no longer an alien force, but indicates an inherent aspect of life. In doing so, we seek to transition away from conceptualizing power solely as the supremacy of social structures over individual humans or the ascendancy of machines over human agency. Instead, our aim is to explore the ascendancy of the natural machinery over the communal fabric of human social life. The dichotomy no longer exists between living labor and alienating machinery, but rather between the living machine and the dead centralized structure. This leads to a universalist perspective in which the distinction between working time and free time dissolves into a singular economy, and machines are no longer confined to factories. Social life as a whole becomes machinic, meaning it becomes the site of political power's operation. Therefore, our goal should be to seek the emancipatory force by reformulating this entire machinery. The focus has shifted from viewing machines solely as tools, to recognizing the inherent machinic essence of social existence—the very structure of the social machine itself—as the subject of critique and reconstruction.

Jacques Derrida in his celebrated text, "Structure, Sign, and Play in the Discourse of the Human Sciences", already attempted to decentralize the concept of structure. He liberates the concept of structure from mechanism and formalism, indicating that the decentralization of structure and its materialization are one and the same act. Undoubtedly, Derrida's thought is aligned with machinic ontology, particularly evident in his critique of logocentrism. It is *logos* that centralizes and de-machinizes any structure. Derrida's "logic of supplementarity", which reflects his attempt to formulate a logic without *logos*, indeed moves toward a guerrilla logic.

The centralized image of the brain as the leading and commanding organ replicates the same image of the *logos*, and therefore is as well misleading. In this context, machinism is an effort to philosophize the human, social and natural structurality without any recourse to center, *logos*, or any commanding organ or any organism.

The idea of mechanical, which is to say, predictable, purposed, and mono-functional machines is only a projection of centralized conscious mind that gives a false image of human body and thought. Mechanical machines are not true machines because they are based on the image of subjective centrality, or say, logocentrism. The same holds for organism. Both mechanism and organism are subjectivism projected on machines and living beings. Machines are not tools; they are automaton. They are endowed with freedom, but their freedom, far from the idea of conscious free will, should be understood in terms of contingency, which is the element of nature. The source of freedom is at the most elementary pieces of nature, as Epicureans with the notion of *clinamen* implied. This idea posits that atoms of nature are not determined by natural laws, but rather possess a slight swerve or inclination that allows for spontaneous and unpredictable movements. This randomness at the atomic level is the source of the contingency and indeterminacy that, according to Epicureans, characterizes the natural world and makes human freedom possible. By recognizing this underlying freedom, we can move beyond deterministic models of nature and embrace a more open and dynamic understanding of the whole universe. A machinic perspective sees everything as a machine composed of natural elements, which are themselves contingent and therefore free. This means that all entities, from the smallest particles to the largest social systems, are shaped by a complex web of factors that are constantly interacting and changing. In this sense, we are beyond the Cartesian or Kantian dualism of human mind and the rest of nature. Schelling's great discovery is that freedom does not stem from human essence but from nature itself. This becomes clear when we consider human societies rather than individual humans. By doing so, we can describe the behavior of a human society in a way similar to that of a society of animals or bacteria. The philosophical question then arises: in what sense can freedom be ascribed to a society? If we place freedom in the minds of individuals, in the Cartesian or Kantian sense, then society itself would become mechanical, just as the rest of nature would. The laws of societies would be similar to the laws of nature, and social sciences would be similar to natural sciences. In reference to Canguilhem and Simondon, we will propose a reformulation of our understanding of natural sciences in terms of the inclusion of a kind of Schellingian element of freedom, and thereby reformulate the social sciences. The conclusion is that, if the element of freedom is locatable in nature itself, it is society that is truly free, not the individual. The freedom of a society is not derivable from the conscious freedom of its individuals.

The pre-individual sociality can be described as a social unconscious. The concept of the machinic unconscious, introduced by Félix Guattari and Gilles Deleuze, serves as a counterpoint to psychoanalytic approaches to subjectivity and intersubjectivity. Jacques Lacan elaborated on the social nature of the unconscious by conceptualizing it as being *structured* like a language. Is Lacan's unconscious the pre-individual social element we are seeking? In part, yes. However, this aspect of Lacan's work remains underdeveloped and somewhat enigmatic. It is often referred to as *objet petit a*. Yet, much of Lacan's structuralist framework undermines this discovery by presenting an unconscious whose sociality mirrors that of conscious individuals. In short, Lacan's emphasis on language and intersubjectivity leads his formulation into logocentrism, even though he frequently oscillates between these two poles. At times, he underscores the materiality and unspeakable essence of words, as seen in his reference to Harpo Marx in "Seminar VII: Ethics of Psychoanalysis". Thus, rather than Lacan's structuralist view, which frames the sociality of the unconscious in terms of intersubjectivity, we should turn to another direction in his work that transcends any account of subjectivity or intersubjectivity. This is the direction pursued by Guattari in "Machine and Structure" under the name of the "*objet-machine petit 'a*'", leading to the concept of machinic unconscious (Guattari 1984, 115). It is crucial to emphasize that there is no personal unconscious; what is often referred to as the personal unconscious is, in fact, a form of consciousness. The unconscious should be understood as inherently social and machinic. Any account of the unconscious that reduces it to a product of structuralist intersubjectivity falls into the trap of treating it as merely an aggregation of individual consciousnesses.

In this regard, our study seeks to distinguish the machine as a philosophical concept from the notions of form and structure. Machine and structure constitute different types of multiplicities. The emergence of logos is equivalent to the solidification of forms from machinic fluid junctions and syntheses. On this basis, we can differentiate between two types of abstraction: machinic abstraction and formal abstraction. Logos is produced out of the formalization of the conjunctions of machinic flows. This holds to the Hegelian view that "to him who looks on the world rationally, the world looks back rationally". It is not only in humans that emotions and drives take a sedimented form and become stabilized collectively and produce reason. Everywhere in nature we find the conjunctions of flows and the sedimentation of collective forces which lead to the emergence of a kind of reason. *Logos* is not our common differentia (*diaphora*). But as long as it is the matter of human knowledge, there is a tendency in philosophy, and even in sciences, to build a closed structure of inter-related ideas. Let's refer to it as ideology for a moment. To avoid falling into ideology in this sense, it is necessary for philosophy and sciences to let themselves disturbed with the *élan vital*, the fluid and unformulated, non-conceptualized, and non-conceptualizable matter, a machinic *élan vital* which is devoid of any living substance. Hence, machinic abstraction, on the other hand, is nothing less than a declaration of war.

A machine does not have parts or organs, but rather flows, or atomic swerves without atoms, the swerves that come before atoms. It does not have a center or a commander in chief. There are many directions in European philosophy after Kant that focused on demonstrating that humans cannot be instruments. Society is a machine, the whole is a machine, but without an end or utility. Althusser's notion of aleatory materialism is illuminating here. The so-called formalism is based on the systematization of natural machines, where there is a structure at the cost of the flow, reducing everything to the management and instrumentalization of society. The result is a hierarchy in which the unity of the system (the manager) has a higher position than the parts that are mere instruments. Our concept of the machine here is in opposition to this instrumentalization and bureaucratic systematization. The machine is the opposite of a tool.

In this sense, the social unconscious is inherently machinic. Contrary to Jacques Lacan, Deleuze and Guattari argue that the unconscious is not a structure, but rather a machine. This does not refute the claim that the unconscious is linguistic; rather, it provides a machinic account of language itself. The source and power of language is the automatic energy that recognizes and reflects sounds and voices. Sense is a product, but it is present from the first reflections of the infant. Structural meaning is a latecomer in this process, against which and prior to it, we can consider machinic sense. We don't only think through pre-existing meanings. We begin to think from the moment of birth, and we are not the only thinkers in the world.

A Review of Existing Literature

Several published works have explored various aspects of machinic ontology. In this section of the introduction, we aim to briefly examine these works, highlighting their strengths and weaknesses from our perspective. We argue that the root of their insufficiency lies in the political underpinnings hidden within these versions of machinic ontology. Thus, after reviewing them, we will conclude this introduction by briefly addressing the political orientation of the current project. All these versions align with the trend known as speculative realism. In the final political section of the introduction, we introduce the concept of machinic abstraction, which implies that our interpretation of machinic ontology represents a materialist idealism. In the conclusion of the book, we will expand on this, defining ideas as dreams.

Jane Bennett's Vital Materialism and its Limits

Jane Bennett's *Vibrant Matter: A Political Ecology of Things* seeks to ground politics in things themselves, rather than human concerns. She argues that this can be achieved by rejecting the mechanistic model of reality in favor of a contingent perspective. In this regard, our project aligns with her approach. In the fifth chapter of her book titled "Neither Vitalism nor Mechanism," she introduces "an impersonal agency integral to materiality as such, a vitality distinct from human or divine purposiveness" (Bennett 2010, 69). After examining Kant, Driesch, and Bergson's ideas of a non-mechanistic material force that exists exclusively in organisms, Bennett seeks to integrate this force into the concept of materiality, reformulating it to encompass both organic and inorganic matter through this force. In this way, she overcomes the traditional divide between mechanistic matter and organic living matter. She shares the belief of Kant and Driesch that the living force (*Bildungstrieb* for Kant and Entelechy for Driesch) is a material force and not a psychic element or soul. However, she rejects the idea that this force exists only in certain material compositions, namely living organisms. Interestingly for our purpose here, she quotes Michael Bakhtin who envisioned "a relentlessly self-constructing, developing machine [which] … builds itself not from pre-prepared parts, but from self-constructing ones" (75–76). In her work, Bennett does not further develop the rethinking of the notion of the machine, "the possibility of a creatively self-organizing or intelligently adaptive machine" (76). Instead, she focuses on her version of vital materialism, which is grounded on the idea that things can disobey our commands.

Bennett's concept of agentic assemblage seems to align with the notion of machinic agency. The fact that things disobey our commands and expectations, as she explains through the example of the north American black out, would ascribe an agency to those objects. However, she never tries to discuss the nature of this agency of things or objects. Furthermore, she understands things or objects in their everyday sense, and her emphasis on their withdrawal does not alter their nature, though it points to something beyond the grasp of human natural sciences. Hence, rather than developing her ideas around the concept of "machinic assemblages", she relies primarily on the notion of vitality as the central philosophical tool for articulating her materialism. Objects are in an unknown way vital, and their vitality is distinct from biological vitality. The problem is that her work does not construct an ontology and cannot extend beyond an invitation to be attentive and respectful towards things and their effects and lives. And these things are litter ("Glove, pollen, rat, cap, stick"; p. 4), electrical power grid, ingested food, stem cells, etc. Despite her initial aim, Bennett focuses on the values and effects of these things on human life. In *Vibrant Matter*, the vitality of things is understood in terms of their interconnectedness with humans, even though these things withdraw from the ways we aim to connect to them. In other words, rather than making objects subordinate to us, we become subordinate to them. Or, rather than the capitalist instrumentalization of things, she advocates for a poetic (and moral) relationship with them. However, subordination and enslavements are also human relationships. The following sentence that she provides in her preface to the book succinctly summarizes the core of her vital materialism: "We need to cultivate a bit of anthropomorphism – the idea that human agency has some echoes in nonhuman nature – to counter the narcissism of humans in charge of the world" (xvi).

The problem with Bennett's project lies in her starting point: she begins with things and their powers. The failed ontological approach leads her to assume the substantial integrity of namable things. Thus, the power of things is their ability to maintain their substantiality, and this power is equivalent to Driesch's Entelechy or Bergson's *élan vital*, which represents the vital force present in every entity. This force is what enables things to resist being reduced to mere relationality. The issue is Kantian: nameable things are human constructs, shaped by the harmony of perceptual faculties. If Bennett aims to explore things in themselves and their agency, she must look beyond familiar objects like litter or digested food. She begins with ordinary things and ascribes extraordinary power to them. For her, things in themselves are just ordinary things considered in themselves. This is not a step toward ontology.

Although Bennett draws the concept of assemblage from Deleuze and Guattari's *A Thousand Plateaus*, she deviates from their non-vitalist understanding. Instead of providing an ontology of assemblages, in the manner of Deleuze and Guattari, she discusses “a power blackout that affected 50 million people in North America in 2003” as an example of “the distributive agency” and “the electrical power grid as an agentic assemblage” (21). We agree that the blackout is a compelling example of distributive agency (a machinic power-structure), but we disagree that the electrical power grid, as described by Bennett, is a good example of an agentic assemblage. As will be discussed in Chap. 5, for Deleuze and Guattari, the notions of assemblage and machine are synonyms and underlie an open articulation of flows. Bennett, on the one hand, understands the notion of machine exclusively in mechanistic and instrumental sense, and on the other hand, considers the vital forces of materialities as the elements of an assemblage. These materialities, in the example of the electrical power grid which, according to her, “offers a good example of an assemblage”, are “electrons, trees, wind, fire, electromagnetic fields” (24). This is the difference between her account of assemblage and Deleuze and Guattari’s machinic account: a difference between machinic flows and the objective materialities and their vital forces. It is not surprising, then, that she favors anthropomorphic accounts of the event over technical ones and argues that ‘anthropomorphizing has…its virtues’ (25).

On the contrary, a machinic ontology would posit that reformulating *élan vital* or entelechy in terms of the agency of assemblages does not necessitate anthropomorphism or reliance on vitalism. It asserts, instead, that all assemblages are machines. The approach should move in the opposite direction, freeing philosophical language from anthropomorphic and vitalist terminologies, liberating the concept of machine from all instrumentalization and teleology, and making it the primary concept of philosophical ontology. Bennett’s description of the blackout event is too concrete to lead to a true materialism. A real abstraction, one that moves beyond ordinary things and human concerns, is necessary to approach real materialism.

Levi Bryant's Machine-Oriented Ontology and Its Limits

The guidelines of a machinic ontology are most explicitly outlined in Deleuze and Guattari's volumes, *Capitalism and Schizophrenia.* Deleuze and Guattari attempt to provide a complicated system that explains many aspects of human psychic and social life, as well as natural phenomena, in terms of desiring machines or abstract machines. However, their rich endeavor is far from being argumentative or straightforward, which makes their work open to further development and expansion. This is the direction we are exploring in the current book.

One of the first efforts to expand on the claim that "everything is a machine" is Levi Bryant's *Onto-Cartography: An Ontology of Machines and Media* (2014). Similar to Jane Bennett's project, Bryant's main aim in this work is to approach a non-mechanistic materialism. However, unlike Bennett, he employs the concept of the machine to achieve this materialism. For Bryant, a machine is "something that operates", and he provides the following definition: "*A machine is a system of operations that perform transformations on inputs thereby producing outputs*" (Bryant 2014, 38). As this definition implies, it seems that the notion of the machine comes to replace that of the object to help philosophy get rid of the qualities and properties of things. Instead of taking them as what they are, an ontology of machines would deal with how they operate, function, or produce. Machines are supposed to be different from objects or things that are classically considered passive and containing certain fixed qualities. This is supposed to be a new ontology whose objects are essentially different from customary objects. Apparently, this is the reason behind the introduction of the concept of the machine into ontology. We believe that Bryant falls short of his aim, as his book elaborates on Deleuze and Guattari's guidelines in a more formal rather than machinic direction.

The core of the issue lies in the logic of Bryant's ontological approach. With a move that betrays his initial definition, he categorizes machines based on the types of their input flows, the types of operations they apply, and therefore the types of their output flows. This leads to a reduction of machines back into the classical category of qualities and properties, but this time in terms of the operations that machines "possess", namely "powers" as "real or actual features of a machine". Everything is a machine, machines are different and there are different categories and types of machines and we can classify them on the basis of their powers and features (41). This is the same Aristotelian logic that is applied to things that are hardly different from passive, non-operational essences. The logic of classifications and categorizations does not help shift philosophy towards a new materialism. The problem with Bryant's study is that he pursues a machinic ontology while using an essentialist, non-machinic logic. By insisting that "machines possess their powers", he limits the machinic ontology to relatively fixed machines that are distinguished from each other based on their properties. If machines are supposed to be functional and productive, they would go beyond such categorization and the acquisition of fixed properties. Bryant's machines, with their unidirectional inputs and outputs, are more like the instrumental (and instrumentalizing) machines in Charlie Chaplin's *Modern Times* (1936), which conceptualizes the machine in a classical Marxist sense, rather than an ontological perspective.

A true machinic ontology requires a logic of pure multiplicities. Bryant, however, persists in using the language of qualities, properties, and numerical and extensive distinctions between machines, although he acknowledges that these are not inherent to machines but rather the outcomes of their operations. Even with this awareness, he has not refrained from identifying and unifying a machine with its intrinsic power, which is what it is capable of doing and what distinguishes it from another machine that has different capabilities. In this way, Bryant objectifies machines, although this objectification does not result in a direct reduction to the subject (i.e., things really do have those qualities in themselves in the absence of any subject). However, this does not truly change the onto-logic, nor does it effect a radical revolution in the way we see things. It does not amount to a genuine machinic ontology, but rather amounts to the traditional ontology of familiar objects rebranded as "machines".

Levi Bryant regards the concept of machine not as a new philosophical concept, but rather employs the term machine mostly in its ordinary and practical sense. He assumes that there are both organic and inorganic machines, which differ based on their distinct properties. Thus, he confuses the philosophical (the transcendental) and the practical (the empirical) uses of the term machine. This is why he claims that machines are structurally open and operationally closed:

> it is not the case that machines are open to *all* flows available in the world. Rather, machines are only open to a small subset of existing flows. This subset will differ from machine to machine. As a consequence, it is not the case that everything can relate to everything else, nor that everything is related to everything else. This is for the simple reason that machines are not structurally open to all possible flows issuing from other machines. (54)

This is an attempt to protect machines from relationality and to preserve their integrity. Passive objects are replaced by functional machines but this functionality does not make machines to destruct their own integrity and the integrity of the common-sensical world. For Bryant, the machine is not at all abstract in Deleuze and Guattari's sense.

In this regard, another shortcoming of Bryant's analysis is the neglect of the historical context in describing machines and the relationships among machines, media, and nature. Bryant writes, "This seems to be a feature of many incorporeal machines such as mathematical equations that are not changed or modified in any way by the inputs that pass through them as they undertake operations. The Pythagorean theorem remains the same regardless of whether or not the right triangle is this or that size, this or that color, or made of this or that material" (51). The transition from Pythagorean geometry to modern geometry (such as Riemannian geometry) is the outcome of gradual changes, as many right triangles affected the position of the theorem in relation to others, ultimately leading to a mathematical revolution. As Albert Lautman aptly demonstrated, an historical approach is essential if we seek to move from mathematics and mathematical theories to metamathematics as a metaphysical perspective on mathematics (Lautman 2011, 89–90, 205). In the same way, only an historical approach can guide us from various ordinary machines towards a metamachinism, a metaphysical concept of machines in which the machine is viewed as a philosophical abstraction. It is precisely this historical approach that makes the works of Canguilhem, Simondon, and Deleuze and Guattari particularly relevant to our study.

This leads us to offer a critique of the phenomenological approach which is generally present in speculative realism. We are not presenting here a phenomenology of machines, as our aim is not to explore how machines can sense or perceive the world. Instead, we will attempt to develop a metaphysics or ontology of machines. The issue with such a phenomenology that appears in the third chapter of Bryant's book in reference to Luhmann and Uexküll as well as in Kristensen's *La machine sensible* is that it is confined to the domain of human conceptions of sensation or perception. In other words, it humanizes or animalizes the machines. Our approach, on the other hand, moves in the opposite direction. It seeks to de-vitalize, de-organize, and machinize life itself to achieve a pure machinic ontology that surpasses the constraints of human-animal issues. This is why we insist that machinic ontology is a materialist metaphysics that goes beyond the analysis of objects within perception.

Arjen Kleinherenbrink's *Against Continuity* and Its Limits

Arjen Kleinherenbrink, in his *Against Continuity: Gilles Deleuze's Speculative Realism* (2018), provides another account of the machinic ontology, of "the machine thesis" which posits that "all entities are machines", again focusing on the work of Deleuze (and Guattari). He considers Deleuze's "machine ontology" reconcilable with the so-called speculative realism, and maintains that his philosophy implies an object-oriented ontology: "it places volcanoes on the same ontological footing as fleeting thoughts, Genghis Kahn, neutrons, and office chairs" (Kleinherenbrink 2019, 2). By this, he means that Deleuze's philosophy does not reduce things to subjective categories or mere illusions. However, Kleinherenbrink's interpretation of Deleuze is reductionist on another level. In his understanding, if everything is a machine, everything has the same ontic dignity, stands at a unique ontological plateau, and this would be a plane ontology of objects. Kleinherenbrink (assumably following Markus Gabriel, Graham Harman and other List-Oriented Philosophers) indicates an obsession to *list* "everything". To these philosophers, the ontic reality is the set of everything that can be enumerated and named. This approach objectifies everything, resulting in a reduction of heterogeneity to a mere extensive heterogeneity of objects. Our work challenges these views by moving away from the common-sensical language of namable objects and, instead, grounding its foundations in intensive heterogeneities. Machines are not ordinary objects like unicorns or the state of Bavaria.

The proponents of object-oriented ontology, including figures like Graham Harman and Kleinherenbrink, reject the reduction of objects into subcomponents, such as neurons, synapses, or elementary particles, as well as larger structures like social networks or power relations. They argue that an office chair, a hurricane, or a passing thought cannot be reduced to any scientific or social construct, but are real as they are. The things in themselves are ordinary things considered in themselves, independent of relations and qualities. Each individual object possesses inherent value, and its distinctiveness and essence should be acknowledged and safeguarded. This is a poetic thought (in a limited sense) that cherishes the distinctiveness of each individual object. However, it is also a defensive tactic aimed at protecting common sense from critical thought and, in particular, remains politically conservative. As Thomas Lemke declares, "In claiming to go 'back' to the object to explore its inner depth, protagonists of OOO cherish a naïve encounter with objects that reintroduces the most subjective values and arbitrary criteria" (Lemke 2021, 22; see also p. 36). Their opposition to the Kantian critique should also be seen as a conservative stance against Nietzsche's reformulation of the Kantian critique through the idea of "philosophizing with a hammer". Object-oriented philosophers cannot protect objects from the critical hammer.

Graham Harman presents the intriguing idea that aesthetics holds the position of "first philosophy" and takes priority over epistemology. (Harman 2007). This notion parallels Emmanuel Levinas' formulation, which positions ethics as first philosophy. This comparison can help us to understand if objects are capable to build an irreducible ontology. According to us, objects lack the capacity to play a role similar to Levinas' concept of the Face. In being objects, they elude singularity and the weirdness that Harman ascribes to them. Unless Harman makes it clear that the objects he refers to are not ordinary known objects; but the examples and descriptions that he provides don't go anywhere outside the domain of the ordinary known objects. What is more, for Kant, the thing in itself cannot be known but can be thought. For the speculative realists, it can be speculated. A philosopher like Deleuze distinguishes between thinking and recognition, and claims that the objects can only be recognized. In contrast, thinking takes place outside of the subject and its correlated objects. Now, by insisting on the objects (and as their examples show, they mean ordinary listable objects), the speculative realists like Kleinherenbrink, Harman or Meillassoux fail in liberating the speculation from the domain of recognition. In this regard, we see no difference between Harman's claim to "enweirding" the object and what modern science, such as quantum physics or chemistry, does with ordinary objects. The only difference is that in the case of modern science, this enweirdment finds more elaboration (Taylor 2016, 210). The elementary particles such as neutrinos or fermions behave differently from objects, and by trying to describe such behaviors, modern physics attempts to get beyond subject-oriented descriptions of objects.

Our project revolves around what Bachelard refers to as scientific abstraction, while integrating it with political abstraction. This involves continually destructing the language of common sense and the state of affairs, including entrenched scientific terms like "objects" or "particles". In this context, machines are abstract; they transform everything into a system of fluxes. This is what Kleinherenbrink neglects in his reading of Deleuze's (and Guattari's) ontology of machines. A machinic ontology does not respect the substantial rigidity of objects and things; it reassembles them into infinitely many possible configurations of flows that are open to new modes of functioning. For Object-Oriented Ontology, a fleeting reverie is irreducible and must be respected as a thing in itself. A machinic ontology, on the other hand, is an idealism that concerns alternatives, rather than the state of affairs. For the first group, an object is irreducible both to its components and to its overarching systems. For us, although a machine is not reducible, it can be transformed into other machines.

Kleinherenbrink is correct in asserting that Deleuze is a philosopher of discreteness, but he is incorrect in suggesting that Deleuze is a thinker of discrete, listable ontic objects. In our opinion, the primary issue with Kleinherenbrink's interpretation of Deleuze is that he conflates the concepts of singularity and individuality, as well as intensive discontinuity with extensive discontinuity. The former requires a series-based model, while the latter can be described through set theory. Therefore, Deleuze's concept of discontinuity does not lie between things, but rather traverses them, dismantling them without subjectively reducing them. Kleinherenbrink divides the virtual realm (of intensities) and discrete objects into two separate domains, asserting that only the latter possess discontinuity: "Everything is not a machine if the virtual realm exists, because then every machine is a mere representation or expression of something decidedly nonmachinic (continuous instead of discontinuous, intensive instead of extensive, and so on)" (Kleinherenbrink, 36). The so-called virtual realm is not a continuous entity; rather, it consists of heterogeneities. Intensities are heterogeneities or multiplicities. The key Deleuzian distinction is not between continuity and discontinuity, or multiplicity and unity, but between two types of multiplicities: one machinic and the other static, one revolutionary and the other conservative. This is evident even in *Bergsonism*, the primary text in which Deleuze elaborates on the virtual. In this work, Deleuze differentiates between extensive and intensive multiplicities, which are both characterized by discontinuities (Deleuze 1991, 39–40). In Deleuze's philosophy, the key distinction lies between the discreteness of objects, which result from subjective categorical differences, and the discreteness of the real, which encompasses the power of things to construct and connect in order to create. The fourfold-ness that Kleinherenbrink underlines (Kleinherenbrink, 38–40) is an instrument for Deleuze to explore the virtual multiplicities associated with each generated object and how they are experienced. Through this approach, the transcendental is no longer just a formal condition of possibility, but rather an actual condition that is present in every singular experience. This implies that empirical objects are products whose conditions of production can be traced back to material-machinic singularities that contain a serialization of flows.

Audronė Žukauskaitė's *Organism-Oriented Ontology* and Its Limits

Why machinic ontology and not an organism-oriented ontology? Why do we insist that the natural machine, which is managed to machinize life itself, is distinct from an organism? In the course of this study, we will argue that an acephalic organism is no longer an organism but a living machine. It is only in this sense that we can approach an ontology. This is what distinguishes our project from Audronė Žukauskaitė's *Organism-Oriented Ontology* (2023), which aims to introduce a revised concept of organism for constructing an ontology. To achieve this revision, she examines various traditions of organicism and organology, which, in our opinion, do not adequately lay the foundation for an ontology. While organicism throughout its history sought to reconcile the contention between mechanism and vitalism by striving to achieve an open system that sustains itself through interaction with the environment, it remains confined within the concept of self-organization. It takes the ability of organisms to self-repair as granted, without adding any explanatory ground to it. We believe this would ascribe an irreducible unity to the concept of the organism, positioning it in opposition to machines, which are characterized by inherent multiplicity. On this basis, throughout her work, Žukauskaitė builds a dualism between organic and inorganic entities. Dualism is a false pluralism. No dualism is possible without a substantial unity of each term. Therefore, organicism is unsuitable for constructing a pluralist ontology.

There are new attempts to revise organicism, giving an account of it which is based on a conception of organism which is not closed but open to the environment. But even there, the organisms are open only in a limited sense. In order to remain an organism, they have to be independent, self-preserving individuals, never capable of forming a true social-historical ontology. Within the domain of organicism, a community or assemblage of organisms is not itself an organism. No society can be organic in the authentic sense. In other words, organicism amounts to vitalism and not a Thanatism, because the death of an organism plays no role in the constitution of a larger organism or nature. Better said, it holds an essential distinction between the inside and the outside, and therefore cannot evade the concept of a center.

Žukauskaitė's Organism-Oriented Ontology aims to reframe organicism by incorporating the concepts of processuality, multiplicity, and potentiality, thus rendering it suitable for ontology. While she attempts to decentralize the notion of organism, she never quite manages to bridge the dualism between the living and non-living, consequently staying at the centralized level of organicity. She chooses to retain the concept of an organism but reinterprets it within the framework of the notion of assemblage. However, following the perspective of Deleuze and Guattari, no assemblage is organic; instead, it is inherently machinic. An organism seen as an assemblage is essentially a non-mechanical machine. Even in her conclusion, when she references Manuel DeLanda's statement, "Conceiving an organism as an assemblage implies that despite the tight integration between its component organs, the relations between them are not logically necessary but only contingently obligatory", she overlooks the fact that this represents the most severe critique of organicism that DeLanda is putting forth (DeLanda 2006, 11–12 in Žukauskaitė 2023, 154). Therefore, in contrast to Žukauskaitė, our claim here is that the only reformulation that leads to ontology involves relinquishing the notion of organism, thereby creating room for the concept of machine and machinic assemblages. Machinic ontology is far from dualistic; it rejects any appeal to a center or a head.

Žukauskaitė claims, "An organism-oriented ontology pursues the Kantian idea that some characteristics defining living beings – organisms – might be helpful to reconceptualize ontology" (Žukauskaitė 2023, 11). Once again, we must emphasize that no ontology can be dualistic. As Deleuze notes, ontology is synonymous with univocity. Consequently, any organism-oriented ontology should assert that *everything* is an organism. However, in her book, Žukauskaitė focuses on the "interactions… between organic and inorganic, or interactions between living and non-living systems on a planetary scale", rather than on incorporating these systems into an ontological framework (14). Therefore, if her work encompasses an ontology, which implies a distinct criterion for comprehending the entire realm of existence, she should elucidate how the concept of organism (albeit in a revised, decentralized form) can serve as such a criterion. How can this revised notion of organism establish the foundation for continuity between living and non-living systems?

For instance, in her chapter on Simondon, she explains the distinction between biological individuation and physical individuation but does not address the transductive unity that, in Simondon's thought, governs both physical and biological individuation within the physical paradigm. In other words, rather than starting from the physical paradigm, she places the biological at the forefront of ontogenesis. While for Simondon, ontogenesis does not permit a metaphysical distinction between the living and the non-living, Žukauskaitė derives from the theory of individuation the foundational role of the organism in constructing an ontology—an endeavor that, in our view, is ultimately unattainable. This is not reconcilable with her initial aim of reconstructing the organism in order to decentralize it, when she writes: "an organism is understood as a process of folding and also as a folded multiplicity composed of heterogeneous parts" (37).

Similarly, in her chapter discussing Catherine Malabou's concept of plasticity, Žukauskaitė interprets it as a quality inherent in living organisms. She views it as "the capacity of the living being to receive form and to give form, and also the capacity to explode form" (77). However, our forthcoming chapter on Malabou will reveal that this interpretation has its limitations. "The realm of the living in general" extends beyond the boundaries of known living organisms and can encompass all non-mechanical or plastic machines. Plasticity in Malabou is not a conceptual tool for distinguishing between plastic organisms and rigid objects. Therefore, Žukauskaitė's entire exploration of the shift from the cerebral (or neuronal) to the mental lacks relevance in her pursuit of ontology since its focus remains confined to the anthropocentric dilemma of human consciousness, or at most, to the realm of living organisms (85–90). By addressing Malabou's concept of "the brain of history," a concept that Žukauskaitė overlooks, we aim to expand the understanding of brain plasticity beyond its characterization as merely an attribute of a living organ. Our goal is to establish an equivalence between the brain and history (deep history), as well as between plasticity and historicity. Additionally, in the concluding section of our chapter on Malabou, we will reference a study by Vazza and Feletti to illustrate how the plasticity of the brain can have cosmological reverberations.

Last but not least, in the second section of her chapter on Deleuze and Guattari, titled "Deconstruction of Organism", Žukauskaitė overlooks the fact that this deconstruction leads Deleuze and Guattari to substitute the concept of machine for that of organism. She insists on retaining the latter concept, albeit in a revised form. She asserts that the notion of an assemblage is equivalent to a revised concept of organism, while overlooking the pivotal role of the concept of machine and machinic assemblages in Deleuze and Guattari's work, and their fundamental divergence from any notion of organism. They introduce the plane of consistency as an alternative to the plane of organism, rather than merely running parallel to it.[2] To address this deficiency and move toward a comprehensive ontology, we dedicate our chapter on Deleuze and Guattari to exploring their concept of the machine. As will be evident in this chapter, in their perspective, life itself is profoundly non-organic, or in our terminology, it can be described as machinic life. Machinic assemblages do not constitute a single category of assemblages (including artifacts or symbiotic systems) running in parallel to organic assemblages. Instead, they embody the essential nature of assemblages themselves.

Therefore, Žukauskaitė's utilization of the terms "organicism" and "organology" can be viewed as a step in the direction of our machinic approach. As she briefly discusses, if organicism entails explaining self-organizing operations without relying on the concept of a vital force, then not only living organisms but also cybernetic systems, or more precisely, machines, can be encompassed within this concept (Žukauskaitė, 96). Hence, it is not the central role of organology in explaining technical machines through the organism model that paves the way for ontology. Instead, it's the reverse direction in which living organisms need to be elucidated through a generalized model of non-mechanical machines

[2] "Thus, stratification and destratification, or the plane of organisation and the plane of consistency, express two tendencies, which run over organisms and bodies" (Žukauskaitė 2023, 67). She quotes a phrase from *A Thousand Plateaus* that hardly supports her view: "The plane of organization is constantly working away at the plane of consistency, always trying to plug the lines of flight, stop or interrupt the movements of deterritorialization, weigh them down, restratify them, reconstitute forms and subjects in a dimension of depth. Conversely, the plane of consistency is constantly extricating itself from the plane of organization" (ATP, Deleuze & Guttari 2004, 297–8).

that are networked but lack organic characteristics. In short, contrary to Žukauskaitė, an "organization" devoid of living substance is precisely what we consider as a machine.

Machinic Ontology: A Political Treatise

Foucault, followed by Deleuze and Guattari, explored various aspects of the social machine and its historical evolution. The so-called societies of control, as introduced by Deleuze, effectively demonstrate the core machinic nature of human societies: just as a machine cannot exist without modulation and free movement, neither can a society. Society is a machine in the sense that it has neither a head nor a center. Our study introduces the concept of the machine as multifunctional, endowed with an inherent multiplicity. We then explore society as a multitude, characterized by a machinic agency that cannot and must not be reduced to the agency of representatives. We will conclude that representative democracy betrays and undermines social agency. In his early works, Foucault, in describing disciplinary societies, emphasizes the nature of power structures and their historical evolution, while overlooking the organic flow of forces in non-institutionalized areas of social life. In our sixth chapter, we will focus on Foucault's later work, connecting his interpretation of ancient Cynicism with Agamben's concept of bare life, examining his critique of democracy, and concluding that a machinic ontology suggests a form of anarchism. Considering society as an anarchic machine raises questions about the structure of power and the issue of governmentality, which we aim to address in the chapter on biopolitics. Our focus will be a distinctive machinic concept of life, drawn from Canguilhem, which we will present as the foundation for our understanding of politics and something that must be defended.

It is important to emphasize from the outset that machinic ontology gives rise to a specific political attitude. The fact that everything is a machine should redefine our understanding of governance, power, resistance, disobedience, and political activism. The political stance of our project emphasizes that machinic ontology is not a mere description of the status quo. Its dynamic nature and the central role of life within it suggest that reality is far from being a state of affairs. Instead, machinic ontology calls for dreaming of alternatives. In the fifth chapter, we discuss, in the context of Deleuze and Guattari, the idea that every machine can be transformed into a war machine, which positions machinic ontology in opposition to any conservative or conformist politics. Machines dismantle state apparatuses. Living machines, composed of flows, are destructive to rigid, mechanical power structures. Replacing the notion of power-structure with that of a machine would significantly alter the concept of power. Our ultimate goal is to arrive at the concept of a megamachine, a global network of power in which resistance and disobedience form the only bonds and connections.

Foucault and Deleuze and Guattari are often criticized for reducing politics to the description of necessary mechanisms, leaving little room for political agency. Building the political implications of this project on the work of Foucault, Deleuze, and Guattari, we aim to counter this accusation by introducing machinic agency in contrast to subjective agency. We address this problem throughout the book through Simondon's concept of relationality and Deleuze and Guattari's notion of desiring production as inseparable from social production, ultimately concluding that agency emerges as the product of tension. Furthermore, Foucault's critique of democracy in *The Courage of Truth* is rooted in his criticism of subjective agency, opening the space for a new form of agency that is no longer individualistic. In this sense, in the conclusion of this study (Chap. 7), we will introduce the concept of communist democracy as an alternative to representative democracy (assigned agency) and the pure anarchy of no matter who rules (absolute subordination). The term "communist" emerges in response to a context where capitalism presents itself as the ultimate realism of "there is no alternative", as Mark Fisher discusses in *Capitalist Realism*,[3] and to capitalist reduction of everyone to either

[3] Mark Fisher begins his book with a chapter whose title, "It's easier to imagine the end of the world than the end of capitalism", refers to Margaret Thatcher's famous saying, and is discussed by Fredric Jameson and Slavoj Žižek among others. Fisher explains this thought provocatively: "That slogan captures precisely what I mean by 'capitalist realism': the widespread sense that not only is capitalism the only viable political and economic system, but also that it is now impossible even to *imagine* a coherent alternative to it" (Fisher 2009, 2). To us the term communism with its idealist resonance, in contrast to the term socialism which has more a realist sense, provides us with opposition to that capitalist realism.

fascinated or fascinating individuals—slaves in search of celebrities, and celebrities in search of slaves (late fascism). In this sense, machinic ontology is a form of political idealism: envisioning alternatives where there is no alternative. Dreaming is the highest degree of thinking.

Roberto Esposito in *Bios: Biopolitics and Philosophy* explores how individuals protect (or immune) themselves from being dissolved into communities, a process he calls "immunization". He connects the concept of immunity at the political-legal level to the biological sense of the term and concludes, following Foucault, a biopolitics according to which "politics is nothing other than the possibility or the instrument for keeping life alive" (Esposito 2008, 46). In line with the "negative and lethal" sense of immunity, power seeks to objectify and subjectify life. According to Foucault's later work and Deleuze and Guattari, this is accomplished intrinsically through the mechanisms of immunity that protect individuals from the communal storms of life. Thus, life in itself holds political significance, which implies the liberation of individuals from self-preserving immune systems such as identities, nationalisms, possession and so on. Under the neoliberalist regimes, the immune system takes on a globalist, individualist form, and the system of power rapidly adapts to eco-biological circumstances while also becoming increasingly individualizing across the world. This creates a paradoxical mechanism, which according to Mark Fisher and Byung-Chul Han is depressive; a mechanism that forces individuals to confront permanent cultural shocks and global values while also prohibiting them from fully embracing these influences by holding onto their national (or European), familial, or sexual identities.[4] Similarly, the concept of security in the state of peace operates with the same military logic that bounds the free individuals within these identities. By economizing everything through the infinite power of money, capitalism expands a state of war (by other means) to every aspect of human life. Hence, capitalist societies are structured to protect individuals, insulating them from communal influences and safeguarding their lifestyles from contamination by others, particularly immigrants or those who live, believe, or behave differently. In these

[4] Although capitalism is always open to new sexes, but is at the same time conservative to these sexes by insisting on the personal identities of those sex-holders.

societies, the subjects of democracy are individuals defined by their properties: identity, nationality, financial status, and the symbolic value they can exhibit. Social life here is governed primarily by exchange—the exchange of abstract properties and values—aimed at fostering consensus among democratic individuals. In this context, politics must serve as the force that sustains life, freeing living energies from their individual confinements. What alternatives could be imagined to capitalism? Could it take the form of a natural, schizophrenic network of flows where all identities and subjectivities dissolve? Or perhaps it could be an extension of the common into the entire social sphere? Or maybe a society of cyborgs—part natural chaos, part capitalist exchange system? A half communist, half capitalist cyborg?

The machinic ontology is based on the idea that political agency is rooted in a primordial communality that precedes the formation of individual identities. This communality is not constrained by organic limits; instead, it arises from an ontological dimension that transcends the boundaries of biological systems. The liberating force of politics can be understood as life, but in this context, life is not confined to organic organisms; it extends beyond biological systems. Here, life refers to any force that generates novelty, whether it's a stellar system that gives rise to a new star or an ecological system that creates a new organic form. The expansion of the concept of life beyond the organism and the extension of the concept of the machine beyond mere mechanism converge in the *a priori*, ontological communality that defines politics as the resistance of life to mechanization. In this context, sovereignty can be defined as what inertializes the functioning of the machine, like a viscous liquid that fills the parts and components of the system. This dynamism can also be understood as a transition from a form of power rooted in physical dominance over land-workers, characterized as "noisy power", to an abstract (yet real) and structured power over wage-workers. This shift makes the functioning of the social machine dependent on both resisting forces and sovereign forces. This new form of power structure grants individuals the freedom to participate in the labor market, though under what Marx described as "mute compulsion" ("*stummer Zwang*"). The transition from noisy power to mute compulsion runs parallel to a shift in the modes of production—from traditional means of producing surplus value to the optimization of the exploitation of workers through machinery, as asserted by Marx in *Grundrisse*. Furthermore, the digital development of capitalism in the latter half of the twentieth century expands this optimization of exploitation to every single member of society, including administrative employees, free workers, scholars and scientists, the unemployed, and even young children. This actualizes the radical danger that Marx observed in the machinization of production: the situation where man works in the service of the machinery. Human society as a whole becomes a machine, and consequently, any alternative must also be machinic. In this context, the political problem can be framed, using Catherine Malabou's term, as the possibility of resistance within a hyper-flexible capitalism. Our suggestion here is to propose a conception of the social machine that includes resisting forces, political movements, and scattered protests worldwide as its true constituents, thereby rendering every machine a war machine. We argue that these resisting forces constitute the true lifeblood of the social machine. However, the war is not between machines, but within them. In the final chapters on Deleuze and Guattari, as well as Foucault and Agamben, we will explore the crucial connection between social production and desiring-production, leading to an articulation of ethics and politics.

References

Bennett, Jane. *Vibrant Matter: A Political Ecology of Things*. Duke University Press: Durham and London, 2010.

Bryant, Levi. *Onto-Cartography: An Ontology of Machines and Media*. Edinburgh University Press. 2014.

Deleuze, G. (1991). *Bergsonism* (H. Tomlinson & B. Habberjam, Trans.). Zone Books.

Deleuze, Gilles and Guattari, Félix. *A Thousand Plateaus: Capitalism and Schizophrenia*. Trans. Brian Massumi. London and New York: Continuum, 2004.

Esposito, Roberto. *Bios: Biopolitics and Philosophy*. Translated by Timothy Campbell. Minneapolis: University of Minnesota Press, 2008.

Fisher, Mark. *Capitalist Realism: Is There No Alternative?* Zero Books. 2009.

Guattari, Félix. *Molecular Revolution*. Tr. Rosemary Sheed. Penguin, 1984.

Harman, Graham. 2007. "Aesthetics as First Philosophy: Levinas and the Non-Human." *Naked Punch* 9: 21–30.

Kleinherenbrink, Arjen. *Against Continuity: Gilles Deleuze's Speculative Realism*. Edinburgh University Press. 2019.

Lautman, A. (2011). *Mathematics, Ideas and the Physical Real* (S. B. Duffy, Trans.). Continuum.

Lemke, Thomas. *The Government of Things: Foucault and the New Materialisms*. New York University Press, 2021.

Meillassoux, Quentin. *After Finitude: An Essay on the Necessity of Contingency*. Translated by Ray Brassier. Continuum. 2008.

Parsa, Mehdi. "A Reconstruction of Gilles Deleuze's Contribution to the Einstein-Bergson Debate". *Transversal: International Journal for the Historiography of Science*, no. 17 (December 2024): https://doi.org/10.24117/2526-2270.2024.i17.08

Raunig, Gerald. *A Thousand Machines: A Concise philosophy of the Machines as Social Movement*. Translated by Aileen Derieg. Semiotext(e), 2010.

Taylor, Carol A. "Close Encounters of a Critical Kind: A Diffractive Musing In/Between New Material Feminism and Object-Oriented Ontology." *Cultural Studies* 16(2): 201–12. 2016.

Žukauskaitė, Audronė. *Organism-Oriented Ontology*. Edinburgh: Edinburgh University Press, 2023.

2

George Canguilhem and the Monstrous Machine

The existing scholarly studies on Georges Canguilhem lacks a detailed examination of the reconstruction of the concept of machine in his work. Influenced by developments in biology, Canguilhem understood machines other than mechanical tools. He provides rather a biophilosophy of machines. Unlike contemporary interpretations, the present chapter asserts that Canguilhem's theory of machines should be sought in his understanding of organisms rather than in his descriptions of mechanical tools. To understand Canguilhem's theory of machines, one must move beyond the duality of machines and organisms in his work. This is why his analyses of organisms, without recourse to vital substances, provide us with a theory of machines.

In "Canguilhem and the Machine Metaphor in Life Sciences", Océane Fiant addresses the machine metaphor in Canguilhem's work. After discussing Canguilhem's critique of mechanistic views on living machines, Fiant briefly suggests that, "rather than explaining organisms by machines, one must explain machines by organisms. Thus, according to Canguilhem, ethnologists are the closest to the constitution of organology" (Fiant 2018, 158). Unfortunately, Fiant does not pursue this line of thought in her reading of Canguilhem. Instead, she focuses on the ontological difference between machines and organisms, presupposing a dualism that

M. Parsa, *Machinic Ontology*, https://doi.org/10.1007/978-3-031-99561-3_2

needs to be overcome. In their turn, Henning Schmidgen and Yuk Hui deny that Canguilhem makes any contribution to the notion of organicist or cybernetic machine (Schmidgen 2014, 243; Hui 2019). Hui refers to Joseph Woodger as well as Norbert Wiener and Gilbert Simondon, as the first thinkers who, in the midst of the twentieth century, overcame the dualism of mechanism and vitalism, neglecting the role of Canguilhem (Hui 2019, 16–17). Although Hui confirms that Canguilhem was the first thinker to use the term "general organology", he places Canguilhem alongside Bergson and considers him a vitalist (Hui, 26, 149). In his work *Recursivity and Contingency*, Hui, similar to Fiant, assumes a dualism between machines and organisms in Canguilhem's work and understands machines as purely mechanical.

Here, we argue that commentators fail to distinguish between two notions of machines in Canguilhem's work: conventional rigid machines and living machines. Therefore, they missed Canguilhem's important thought that what is essential to life is also essential to technique. The present chapter aims to explore this idea in terms of the central role of milieu. It is true that when Canguilhem compares the characteristics of machines and organisms, as seen in "Machine and Organism", he perceives machines as rigid mechanical entities. All the mentioned commentators adhere to this interpretation. However, this chapter aims to discuss Canguilhem's revisiting of Descartes' non-rigid depiction of machines and his method of understanding both technique and life within the same metabolism, in order to explore the concept of living monstrous machines.

From Descartes' Living Machines to Canguilhem's Epistemological Turn

The story begins with Descartes' attempt to present a mechanistic perspective on all aspects and levels of nature, including the living body. He is credited with coining the machine metaphor to describe the human body. In the *Treatise on Man* (1662 in Leyden), this metaphor aids Descartes in clarifying bodily movements and living characteristics apart from the human soul (Descartes 2004, 99–102). Although the human body is supplemented with a soul, since the soul has no effect on bodily movements, the human body functions like a machine, similar to the animal body, which is unfortunately never supplemented with a soul. A consequence of treating the human and animal body as a machine is the impact it has on the notion of a machine: Descartes introduced the concept of a machine that is not composed of solid parts, but rather of fluxes. In other words, Descartes introduced the concept of a natural machine that is much more complex than manmade machines because it is composed of a multitude of microscopic components. In his *Treatise on Man*, Descartes describes the human body by first examining the flow of heated food in the digestive system. He then goes on to analyze the function of the flow of blood in the body (104–105), the movements of "the machine of the body" (106–108), the nerves and brain (114–120, 142–148), the senses and perceptions (121–125, 149), the eyes (126–135), and so on, viewing all of them as parts, organs, and functions of a machine. Descartes' idea of the machine was inspired by manmade instruments, such as "clocks, artificial fountains, mills, and other similar machines which, even though they are only made by man, have the power to move of their own accord in various ways" (99). However, he draws a philosophical metaphor from these manmade machines to create his own concept of the natural machine. Both the human body and artificial machines share a capacity for self-movement or automation, but differ in the sense that the former is constituted of flows, while the latter is composed of solid pieces. In his work *Principles of Philosophy*, Descartes endeavors to provide an explanation for every phenomenon in nature using the laws of mechanics, while also emphasizing that these phenomena are composed of tiny particles that are not directly perceptible to the senses (Descartes 1982, § 201, p. 284). He compares the natural bodies and artifacts, natural and artificial machines, in terms of the scale of their components:

> when I noticed some similar effects in perceptible things, I judged that these things had been created by similar encounters of such imperceptible bodies; especially when it seemed that no other way of explaining these things could be devised. And, to this end, things made by human skill helped me not a little: for I know of no distinction between these things

> and natural bodies, except that the operations of things made by skill are, for the most part, performed by apparatus large enough to be easily perceived by the senses: for this is necessary so that they can be made by men. On the other hand, however, natural effects almost always depend on some devices so minute that they escape all senses. And there are absolutely no judgments {or rules} in Mechanics which do not also pertain to Physics, of which Mechanics is a part or type. (Descartes 1982, § 203, p. 285)

The qualitative difference in the *Treatise on Man* between solid parts and fluxes is translated in *Principles of Philosophy* into a quantitative difference in the scale of the components. This results in a quasi-qualitative difference between what can be perceived and what is imperceptible to the senses. Despite the differences in composition, both natural and artificial beings share a fundamental essence: they are both bodies, and bodies can be considered machines to the extent that they are self-movable. Descartes concludes that mechanics, as the science of artificial machines, pertains to physics, the science of nature, since both artificial and natural machines are composed of components that move and collide according to the laws of mechanics. This applies not only to inanimate objects, but also to living beings and planetary systems. Thus, for Descartes, everything can be considered a machine, but all machines, even artificial machines, belong to nature and follow the laws of physics. Descartes' view of the natural world as composed of machines led him to conceive of a machine composed of fluxes or molecular components as the proper instrument for describing the entirety of the material world, or nature.

Although Descartes emphasizes the role of mechanics as the science that examines moving bodies, but he opens the way for a revision of mechanics by submitting it to the physics of flows or imperceptible particles. This leads to a materialism, but not necessarily to a mechanistic materialism. Therefore, when Julien Offray de La Mettrie wrote his treatise *L'Homme Machine* in 1747, presumably under the influence of Descartes' *Treatise on Man*, he pursued a materialist direction of Cartesianism, namely non-mechanistic physicalism. When Descartes writes, "nature should … produce its own automatons, which are more splendid than artificial ones – namely, the animals", this can be construed not as a mechanization of animals, but as an indication of the splendid power of physical nature. After all, according to Descartes, nature is the *artifact* of God.

Rightly or not, the Cartesian mechanistic view has been subject to criticism in the romantic or vitalist traditions, starting from Kant's *Critique of the Power of Judgment*, and continuing with the works of Herder, Hegel, and Bergson. It is important to identify the exact target of this critique. These critiques of the mechanistic worldview can be seen as part of a larger movement that challenged the reductionist and mechanistic worldview of the Enlightenment. They rely on the notion of a living force that is present in organisms and distinguishes them from dead and mechanical matter. All of these philosophers held, in their own way, a dualistic view of machines and organisms. They believed that it is not possible to fully explain life and the living world in terms of mechanistic laws. But as Jane Bennett demonstrates in the case of Kant, this is a material dualism, different from the Cartesian duality of matter and soul (Bennett 2010, 71). Descartes divides the spiritual and material realms and subordinates the entirety of the latter to physical laws.

As briefly mentioned, within the realm of the physical world, Descartes is a monist who sees no essential difference between animals and machines. On the other hand, anti-Enlightenment thinkers introduce a dualism within the material world, distinguishing between mechanical matter and organic matter. For both romantics and vitalists, life is a material force that is distinct from psychological elements and cannot be fully measured or explained by mechanistic laws. The romantics and vitalists rejected Descartes' view that organisms can be identified to machines. They argued that living beings possess an animating force that is absent in machines. Furthermore, organisms are not fully determined like machines. But the question is, are machines determined according to Descartes? What these thinkers did not explain is the distinction between a dead body and a functional machine, or more significantly, the operation of a natural machinic system that moves autonomously and is not predetermined.

This is the task that George Canguilhem took on, namely applying the concept of indeterminacy from the life sciences to the physico chemical sciences and drawing conclusions about the idea of natural undetermined machines. He returns to Descartes, by liberating life sciences from any stain of an animating non-physical force. In his early text, *Descartes et la technique* (1937), Canguilhem discovers the importance of Descartes' monism, by asserting that "The radical distinction between soul and body, thought and extension, implies the substantial unity of matter, whatever its form, and thought, whatever its function" (Canguilhem 1994, 227). This represents the profound unity of life and machines, a unity in their functioning, both devoid of soul. This is evident in Descartes' animal-machine theory, where Canguilhem perceives not animals as mechanical lifeless machines, but as living machines, highlighting the machinic nature (soullessness) of life itself. The absolute dualism of body and soul leads Descartes to the intriguing idea that the soul has no effect on bodily movements, and that the body functions with complete independence (231). In this regard, Canguilhem's interpretation of Descartes leads to the notion of the convertibility of life and technics (223–225).[1]

Our aim here is to trace the Cartesian concept of living machines in the work of Georges Canguilhem, striving to articulate his non-mechanistic Cartesianism. Canguilhem is the genuine thinker of technique and life in interrelation with each other. He paves the way for philosophical efforts to put the problem of technique within the problem of life, "rooting technique in life" (Vázquez García and Toscano López 2021, 84; Hacking 1998, 207). To this end, following biologists like Uexküll and Goldstein, he situates the problems of knowledge and life, normativity and abnormality, within the context of various relationships with milieu. This leads to an understanding of life that is rooted in technicity.

Furthermore, this holds considerable implications for the connection between science and philosophy, as well as for the essence of science and technology, because Canguilhem posits that the truth of philosophy (life)

[1] Yuk Hui, in his interpretation of Canguilhem's reading of Descartes, focuses on "Canguilhem's critique of Descartes" and Descartes' "mechanism," neglecting Canguilhem's Cartesianism and his naturalistic materialism, which we emphasize here (see Hui 2019, 153–156). He affirms that "In Descartes there are traces of organological thinking" (157), but this does not help him to revise his understanding of the Cartesian mechanism.

and the truth of science (technology) are inherently unified.[2] In a 1965 interview with Alain Badiou, Canguilhem establishes an inherent link between science and philosophy, asserting that there is no truth except scientific truth and that philosophy has always been the discourse of truth.[3] Here, Canguilhem adopts Bachelard's revision of Comtian positivism, glorifying scientific truth not in opposition but in support of metaphysical abstraction. It is as though Bachelard endorses the scientific mind for its capacity for abstraction, which enables it to detach itself from popular and vulgar knowledge.[4] In this context, Canguilhem focuses his studies on describing life and its capacity to produce knowledge in relation to its milieu. His non-mechanistic Cartesianism closely aligns with his revision of Comtianism, presenting a non-mechanistic positivist monism.[5] Within this framework, there is no room for the vital principle, "vis vitalis"; rather, emphasis is placed on the constructive role of the milieu, as illustrated in Canguilhem's text on Auguste Comte (Canguilhem 1994, 240–242).

To explore the concept of living machines in Canguilhem's work, we aim to converge two paths: first, how he decentralizes, dis-organizes, devitalizes, acephalizes, or, in a word, machinizes the living organism; and second, how he situates the problem of technicity, and thus the notion of the machine, within the context of life. The result would be the monstrous machine as a synthesis of a purely reflexive organism and a physicochemical machine. In this way, instead of "vitalist" theories of reflex theory that Canguilhem criticizes, and "mechanist" theories in nineteenth century, he offers a *machinist* theory of reflex. As Henning Schmidgen observes, Canguilhem's reflex theory provides "the assimilation of two kinds of forms: biological and technological" (Schmidgen 2014, 248). Living machines are not "reflex machines", such as Grey

---

[2] As Canguilhem in *Ideology and Rationality in History of Life Sciences* emphasizes Bachelard's understanding of truth as the core of scientific knowledge (Canguilhem 1988, 10–11).

[3] https://www.youtube.com/watch?v=SKBNSabxhUA.

[4] Bachelard introduces his idea of scientific abstraction in the Foreword of his *The Formation of Scientific Mind* where he claims "We shall have to prove that abstraction clears the mind of encumbrances, that it lightens the mind and makes it more dynamic" (Bachelard 2002, 18).

[5] To see that Canguilhem's Comtianism is revisionary, refer to Elie During's "'A history of problems': Bergson and the French epistemological tradition" (During 2014, 4–23).

Walter's tortoises (243). The problem lays not on the fact that machines are made on the basis of living organisms, but lays on partial automatisms within a living body (Canguilhem 1977, 128). Therefore, the focus on reflex theory should not be considered a mechanization of organisms, but a machinization in the sense that we observed in his reading of Descartes. A theory which is invented by vitalists and then became mechanized, leads to a synthesis between vitalism and mechanism.

Liberating Life from Vitalism; Liberating Machines from Mechanism

Canguilhem's philosophical approach to biology encompasses a vitality that goes beyond vitalism. He endeavors to formulate a philosophy of life that is distinct from, and even contrary to, the philosophies of Schopenhauer, Nietzsche, and Bergson (Talcott 2019, 41). This requires asserting a difference in nature between organisms and dead bodies, without resorting to the notion of a living substance. Accomplishing this is no easy feat, as it involves providing a scientific explanation of life as inherently distinct from dead objects, while refraining from presupposing the existence of a living substance. Canguilhem resolves this dilemma by seeking to elucidate life through the lens of diverse relationships with the milieu. For him, the milieu becomes the true location of life. In this picture, organisms become living machines operating through indeterminate laws of probabilistic sciences.

Canguilhem's aim is not to restrict biology to the study of a particular sphere of existence, namely that of living organisms, but to engage in a philosophical inquiry into biology that has ontological consequences. As Wolfe and Gandolfi observe, Canguilhem is not a philosopher of biology, but what he proposes is a biological philosophy ("*philosophie biologique*") (Wolfe and Gandolfi, 2023, 2). He seeks to identify a distinct approach to the relationship between science and its object within biology and extend it to the broader domain of existence. Biology has the potential to free science from the constraints of mechanism. In this regard, Canguilhem believes that it is the influence of the biological principle that transformed the trajectory of physics and chemistry, shifting them from positivist and objectivist sciences to probabilistic sciences. Therefore, it is not solely biology that defines its object in relation to milieus. In *Knowledge of Life* (*La connaissance de la vie*), a collection of articles published between 1945 and 1962, Canguilhem states: "After three centuries of experimental and mathematical physics, milieu, which in physics first signified environment, has come to signify center- in both physics and biology. It has come to signify what it signified at its very origin. Physics is a science of fields, of milieus" (Canguilhem 2008, 70). Canguilhem's stance is indeed against mechanism in both biology and physics. In biology, he opposes the reduction of life to mechanistic principles of physics and chemistry. Instead, he proposes a return to physics and chemistry to reconfigure them based on the insights gained from life sciences. On the other hand, Canguilhem firmly opposes reducing the subject matter of biology to a distinct domain that arbitrarily transcends the laws of physics and chemistry. In this sense, he endeavors to emancipate biology from dualistic spiritualism (71). He is indeed a staunch materialist. What separates life sciences from physicochemical sciences lays not on believing in a spiritual nature of life. Tom Eyers underlies this point in his reading of Canguilhem:

> If there's an error in classical vitalism for Canguilhem then, it lies in its overemphasis on the *ineffability* or *immateriality* of life. Rather, life permeates organic matter, is in some sense indistinguishable from it, and yet points beyond the limits of inert matter to the ways in which organisms must maintain a relationship to the changing context of their milieu in order to remain alive. (Eyers 2013, 158)

The notion of vitality without vitalism is equivalent to a concept of materialism without mechanism. A non-mechanistic materialism needs to be introduced and clearly distinguished from any form of spiritualism. The major mistake lies in perceiving a living organism within a mechanically understood physicochemical milieu, treating a living entity as a living substance within a dead environment.[6] Life is not an exception

[6] This reading contrasts with Eyers, who seeks to distance himself from "any contextualist understanding that would reduce an organism to the totality of its relations or to the structural background within which it persists" (Eyers 2013, 158). While we do not aim to offer a contextualist perspective by emphasizing the structural nature of the milieu, we do want to highlight that, for

within a scientifically predictable milieu. The factual existence of life alters our comprehension of both the nature of the milieu and the laws of nature. It is only a mechanistic perspective of science that reduces it to mere prediction. However, with Canguilhem, biology and medicine emerge to rectify our understanding of scientific endeavors and our relationship with nature. Particularly in the chapter "Monstrosity and the Monstrous" in *Knowledge of Life*, through recalling a brief history of monsterology and teratogeny, he tries to define life in terms of a pervert *élan* (Canguilhem 2008, 137). If nineteenth century made monsters into scientific objects, deviations from natural development, Canguilhem wants to make nature itself, and its science, monstrous, bringing back to science the glorious thing that had to seek refuge in poetry: "how can we resist the temptation to find the monstrous once again at the very heart of the scientific universe" (143). If mechanics refers to the straight laws of nature, life sciences indicate an essential perversion in nature that generates incessantly monstrous forms, as Canguilhem quotes Isidore Geoffroy Saint-Hilaire: "There are no exceptions to the laws of nature, only exceptions to the laws of naturalists" (142). Monstrosity and perversion belong to the very force of nature and this would be only possible when we regard the relation with milieu in every level as far from a deterministic one. This is why, for Canguilhem, mechanization is identical to rationalization.

In *Knowledge of Life*, Canguilhem establishes a unification between what is traditionally regarded as inherent to nature, namely life, and what is typically perceived as human-made and artificial, namely knowledge. According to Canguilhem, knowledge is generated through the organism's interactions with its milieu. It is the technique that enables the organism to overcome obstacles and challenges. Every organism must generate knowledge to ensure its own preservation. Canguilhem's revolutionary principle is that knowledge is an effect of life, and scientific

Canguilhem, the milieu is the genuine site of life. Eyers is correct in asserting that Canguilhem's thought is far from structuralism or contextualism (or a "metaphysics of virtuality", p. 159, or "relationism", p. 161). However, Eyers does not sufficiently address the central, non-structuralist role of the milieu in his brief reading of Canguilhem. Instead, he repeatedly emphasizes that Canguilhem is not a structuralist or relationist, a point we find puzzling, as we have encountered few, if any, structuralist interpretations of Canguilhem.

knowledge must align itself with this fact. Life and technique precede knowledge and science. Therefore, he declares that "if thought and knowledge are inscribed within life so as to regulate it – as is the case with man – this very life cannot be the blind and stupid mechanical force that one likes to imagine when one contrasts it to thought" (63). This is a straightforward argument: Thought and knowledge are inherently embedded within life; therefore, it is life itself that possesses these capacities. In this manner, Canguilhem unifies man and nature within a single domain of study, not by reducing man to a mechanistic view of nature, but by emphasizing the complexity of life, which enables the emergence of thought and knowledge. Even when he discusses the role of imagination in the generation of monstrosity, he does not build a dualism between an imaginatory monstrosity in humans and a biological monstrosity of animals. Rather, referring to a text from Johann Theodore Eller, "a dissertation that credits animals with the power of engendering noteworthy monstrosities via their imagination", ascribes the monstrous power of imagination and dreams to all animals and organisms (Canguilhem 2008, 139).

Monstrosity is a term that indicates the irrationality of knowledge, contingency of science and non-mechanicity of technology and machines. In the opening pages of the Introduction of *The Knowledge of Life*, Canguilhem expounds upon a distinction between nature and the laws of nature, qualities and numbers, beings and relations (xvii). He endeavors to elucidate not only the mode of existence and the manner of production of the former but also how our understanding of this mode of existence can influence the latter, specifically the essence of what is commonly referred to as quantitative science. When we *know* that our scientific instruments *are* natural effects, this knowledge significantly impacts how we employ our scientific instruments. This is the sole viable approach for attaining scientific knowledge of life. Without this fact (and this facticity), any scientific knowledge of life inevitably leads to its reduction into mere mechanistic explanations. In other words, only living machines are capable to examine life. Hence, Canguilhem employs life sciences to introduce a non-mechanistic perspective towards science as a whole. In this sense, in his work *The Normal and the Pathological*, Canguilhem, making a reference to Claude Bernard, asserts that the scientific realization of observation influencing the observed phenomenon was discovered in the fields of medicine and biology long before it emerged in wave mechanics: "When we glorify the contemporary theorists of wave mechanics for their discovery that observation interferes with the observed phenomenon, it happens that, as in other cases, the idea is a bit older than they are" (Canguilhem 1991, 147).

The Lived Experience of the Sick

In *The Normal and the Pathological*, his doctoral thesis and debut book, Canguilhem orchestrates a substantial relationship between the pathological, or anomaly, and the existence of the organism, paralleling the relationship between the monstrous and life. Intriguing for our purposes is how he understands technicity and industry as the essential anomaly, thereby defining technique in terms of the essential characteristic of life, namely, correspondences with the milieu.

The Normal and the Pathological contains, as Canguilhem clarifies, an "ontological theory of disease" (Canguilhem 1991, 39). According to him, the problem of defining the normal and differentiating it from the pathological serves as the metaphysical condition of all medical treatments. This is because medical interventions aim to heal patients while simultaneously evading a precise definition of what constitutes normality. Can normal be defined ontologically through the concept of a natural state, or dynamically through the notion of an internal equilibrium? Canguilhem introduces a third alternative, rejecting both ideas of natural state and internal equilibrium. His departure point is not defining the normal state and then approaching the definition of disease. Rather, he says, "with Gabriel Tarde, that the normal type is the degree zero of monstrosity" (Canguilhem 2008, 136). In fact, he provides a pathological concept of life. Ultimately, this approach would lead to a redefinition of psychological and sociological concepts of norms based on the medical and biological understanding of the pathological.

Canguilhem deconstructs the traditional perspective that regards the pathological (or the problematic) as simply an imbalance, disorder, or deviation from the natural state of equilibrium. One of the pieces of evidence he presents is that in medicine, we cannot rely solely on nature to restore the desired equilibrium. If that were the case, there would be no need for the invention of technical instruments and chemical medications. Instead, we would simply wait for the organism to naturally restore itself: "if we delegate the task of restoring the diseased organism to the desired norm to technical means, either magical or matter of fact [positive] it is because we expect nothing good from nature itself" (Canguilhem 1991, 40); we must employ technical means to restore the diseased organism to the desired norm. The desired norm extends beyond the mere notion of the natural state. Canguilhem concludes that "disease is not simply disequilibrium or discordance; it is, and perhaps most important, an effort on the part of nature to effect a new equilibrium in man" (Canguilhem 1991, 40).

"... on the part of nature"! There is a paradox to be solved here. Firstly, there is no inherent natural norm or a natural state of stable equilibrium to which any diseased organism must revert. Instead, there are constantly emerging new equilibriums and a natural drive to surpass, enhance, and transcend the current state of normality. In this context, "normal" implies mediocrity rather than excellence, aligning with the etymology of the term, as Canguilhem also alludes to (Canguilhem 1991, 125). But on the other hand, this creative force originates from within life or nature itself: "since life is capable of failures, all its successes are failures avoided" (Canguilhem 2008, 136). The only way to resolve this paradox is to understand the concept of nature not through mechanical laws of nature, but through Canguilhem's understanding of life. Life is not mediocre; it utilizes technical instruments to augment the normal equilibrium, whether in bacterial, animal, or human manifestations. Life is inherently creative and must express its creativity. The creative essence of life serves as the foundation for medicine's technical interventions. In this manner, Canguilhem's biological and medical studies converge and intertwine. Medicine has incited a transformative shift in what is still referred to as "human". However, this technical transformation is not in opposition to nature. The crucial aspect is that nature does not adhere to any fixed norm or stable state; it is fundamentally technical. Canguilhem emphasizes that only life itself, and solely an organism, can be pathological. It is only within an organism that sickness can appear (Canguilhem 1991, 127).

He is drawing here a distinction between organic and inorganic nature. The inorganic nature cannot experience sickness since it remains solely objective to our knowledge and does not entail the kind of knowledge interference that Canguilhem attributes to biology and medicine. However, the fields of wave mechanics, quantum physics, and chemistry have revealed a similar phenomenon in non-biological domains. What Canguilhem uncovers reveals a crucial aspect of science in its entirety. The experimental nature of medicine is intrinsically connected to the experimental nature of life itself. This imparts a contingent element to science as a whole. Therefore, although inorganic objects cannot become sick, if one considers the entirety of nature as an object, life and technicity can be seen as its sickness or monstrosity.

Therefore, diseases carry significant scientific value in terms of comprehending the essence of life and revealing the creative force inherent in nature. Referring to Leriche, Canguilhem states, "Like Leriche we think that health is life in the silence of the organs, that consequently the biologically normal, as we have already said, is revealed only through infractions of the norm and that concrete or scientific awareness of life exists only through disease" (Canguilhem 1991, 118). The same force that generates monstrosities and anomalies in nature is also responsible for generating excellence and miraculous mutations: "disease is not a variation on the dimension of health; it is a new dimension of life" (Canguilhem 1991, 186). Again, in reference to Leriche, Canguilhem claims, "disease is not merely the disappearance of a physiological order, but the appearance of a new vital order" (Canguilhem 1991, 193). Samuel Talcott succinctly encapsulates all these aspects: "Disease is thereby creative, not only insofar as it is a new way of living, but also insofar as it calls up both a technical invention and a will to know" (Talcott 2019, 149). And a few pages later, he maintains, "Disease … is the spontaneous creation of new biological norms, new behaviors, in an altered milieu" (Talcott 2019, 154). The knowledge of life is not the knowledge of healthy life.

In this manner, Canguilhem emphasizes the role of disease in the natural evolution of life. Health and disease signify distinct and essential aspects of organismic existence, and it would be incorrect to define and comprehend the essence of a living being solely based on its state of health. That is why Canguilhem places emphasis on the concept of the pathological in medicine, as it allows for the analysis of the "lived experience of the sick" (Canguilhem 1991, 115). Being sick represents a unique state of existence, one that is highly intriguing.

Similarly, in the third chapter of *The Normal and the Pathological*, Canguilhem elucidates that there is no normal milieu; rather, it is the relationship with the milieu that can be normal or pathological (Canguilhem 1991, 143). In the case of living organisms, there is no fixed or homogeneous space serving as a milieu with which the organism can either establish or fail to establish equilibrium, thus categorizing it as normal or pathological respectively. The categorization of being normal or pathological is relative to the relationship between the organism and the milieu, rather than being determined by the organism or the milieu in isolation. The milieu presents itself as a problem that the organism must confront and improve upon. Normal denotes the absence of such problems, indicating a state in which the relationship with the milieu appears as a peaceful equilibrium. By combining this observation with Canguilhem's assertion that only a whole can be sick, it follows that there is indeed no organism or body that can be regarded as independent of its milieu. The milieu is inherent to the organism, and Canguilhem goes a step further by introducing Bernard's concept of the inner milieu (Canguilhem 2008, 7).

Now, for Canguilhem there is no essential difference between natural milieus and laboratory or industrial milieus. He illustrates this by using the example of the color of butterflies in natural environments compared to industrial regions (Canguilhem 1991, 143). In natural environments, grey butterflies are better camouflaged and have a higher chance of survival by avoiding being preyed upon by birds. However, in industrial regions where bird populations are lower, butterflies can maintain their black coloration without facing significant risks. The distinction between grey and black butterflies as distinct species is a direct consequence of the changes in the milieu or environment they inhabit. Neither the milieu nor the organism is inherently pathological, and any random change in the milieu, known as an anomaly, may or may not lead to a new state of equilibrium. Stuart Elden, in his commentary on Canguilhem, concludes, "Anomaly is not pathological because it is an anomaly, because some of these mutations allow the living being to be normal – that is, adaptive" (Elden 2019, 20). This example serves to demonstrate that industrial regions and man-made environments can operate under similar principles and follow the same laws of interactions between the organism and the milieu. It highlights the interconnectedness of natural and human-made systems and how they can both exhibit similar dynamics and influences on organisms within their respective milieus. If we consider industries as anomalies in comparison to nature, butterflies can establish a normal and healthy equilibrium within these industrial environments. Similarly, the human inhabitants of these industrialized areas create their own norms and adapt to the specific conditions and challenges presented by their surroundings. It is within this circumstance that one should consider Canguilhem's theory of technicity and his thought on machines.

The relativity of norms is equally applicable to scientific experimentation and the laboratory as an anomalous setting. Just as the organism's interaction with its milieu determines what is deemed normal or pathological, the laboratory setting establishes its own set of norms and conditions. Scientific experimentation operates within this constructed framework, wherein deviations from established norms can pave the way for novel discoveries and insights. Therefore, akin to the significance of industrial regions for butterflies, the laboratory can be regarded as anomalous due to its artificial construct. This recognition highlights the role of epistemology in ontology as the realm of equilibriums or beings. Knowledge takes shape when a mutation in the milieu compels specific directions and formations within the organism. This necessitates the separation of the object of knowledge from its natural environment and the establishment of an artificial norm that aligns with cognitive faculties. This principle applies not only to biological objects but also to physical objects. As previously mentioned, Canguilhem draws a comparison between this phenomenon and the discovery of wave mechanics in which "observation interferes with the observed phenomenon" (Canguilhem 2008, 146–7). One can conclude from this that technical objects, when considered not based on their planned usability but in terms of their scientific functionality—which is only understandable within the context of the history of technology—operate under the same laws that govern living organisms.

In this perspective, we can trace the transition that Canguilhem undergoes from biological norms to cultural norms. His viewpoint in this regard is explicitly non-human or anti-human, as he disregards any notion of universality of humans on Earth. Instead, his focus lies on geographical and historical factors, following the archeological and ethnical work of André Leroi-Gourhan. Briefly, human norms are the products of equilibriums with diverse and ever-changing milieus. Similar to how butterflies in industrial regions may be black instead of grey, humans in the industrial age manifest distinct characteristics compared to the agricultural age. Just as transitions and the emergence of new species can occur among animals due to such factors, the same possibility exists for the so-called human civilization. The industrial human can be regarded as a distinct species that emerges out of the agricultural man within the same metabolism that generates more complex biological forms out of the simpler ones.

The Organism as Machine

The recognition of milieu as the center fundamentally transforms the traditional perspective on the integrity of living organisms. Living beings assess their milieu by selectively engaging with non-living elements and resources that maintain equilibrium, thereby shaping the milieu to accommodate their diverse requirements. Hence, photosystems in plants and certain bacteria select a distinct resource (light) from the environment, distinct from systems that absorb water or minerals. Each system establishes its own unique definition of the milieu. The principle of choice in living beings relies on this differentiation between systems, necessitating a form of evaluation. However, this evaluation is not merely passive based on a principle of choice; it actively shapes the constitution of the milieu through interactions. A notable example of this is the "oxygen catastrophe" or "oxygen holocaust", which is previously discussed elsewhere (see Parsa 2019, 27–29).

One of the implications of this line of thinking is the recognition that diverse life forms can coexist within a single organism, and an individual organism can be regarded as an integral component of a larger living system. This implies that a scientific account of an organism cannot be based on an organic substance. In this context, the term "scientific" should be understood in its machinic sense, referring to an approach that emphasizes the analysis of components, connections, organs, and functions. This machinic understanding of organisms arises from the multiplicity of systems or the differentiation of value-systems, which challenges notions of substantiality or individualism.

In his lecture titled "Machine and Organism", Canguilhem critiques the mechanistic approach in biology and introduces the concept of dialectical materialism as a counterpoint (Canguilhem 2008, 75). He defines the "function of the organism" as opposed to "an *already-constructed* machine" (76; emphasis added). In this manner, Canguilhem affirms that organisms are fundamentally distinct from constructed machines. However, this does not imply that they possess a vital substance. Instead, he asserts the notion of explaining machines based on organisms, challenging the mechanistic Cartesian thought that views organisms as machines.[7] He endeavors to redefine the concept of machine based on his novel comprehension of organism. This new understanding indeed involves a dialectical materialist explanation of organisms. According to Canguilhem, the scientist's explanation of the nature of a machine must differ and even oppose the engineer's explanation, which defines its nature in terms of its usage.

Canguilhem considers the mechanistic account of the machine in terms of a manmade arrangement comprising solid components in motion. The crucial and intriguing aspect of any machine is that its configuration remains intact throughout the movements of its parts. At this point, machines and organisms are similar. What sets apart a machine from an organism, according to Canguilhem, is that machines are manmade and lack autonomy. In contrast to machines, organisms possess the capacity for "self-construction, self-conservation, self-regulation, and

[7] Daniel Nicholson in his review of Canguilhem's *Knowledge of Life* summarizes this point as the following: "Since Rene Descartes it has become standard practice in biology to think of the structure and function of organisms on the basis of the structure and function of machines. Only rarely has anyone sought to understand the machine on the basis of the organism. This is what Canguilhem sets out to do. His central claim is that technological artifacts are not so much the product of man's intellectual activity as they are a reflection and an extension of man's own behaviour as a living being" (Nicholson 2012, 2).

self-repair" (88). Moreover, a machine is designed to be monofunctional and strictly defined in its purpose. On the other hand, an organism exhibits much greater polyvalence and can live pathological states. Intriguingly, Canguilhem claims,

> We must at least admit that, in the organism, a plurality of functions can adapt to the singularity of an organ. An organism thus has greater latitude of action than a machine. It has less purpose and more potentialities. The living organism acts in accordance with empiricism, whereas the machine, which is the product of calculation, verifies the norms of calculation, that is, the rational norms of identity, consistency, and predictability. Life, by contrast, is experience, that is to say, improvisation, the utilization of occurrences; it is an attempt in all directions. From this follows a massive and often neglected fact: life tolerates monstrosities. There is no machine monster. There is no mechanical pathology. (89–90)

According to Canguilhem's argument, the notion of a machine monster, which he claims does not actually exist, can serve as a captivating concept for contemplating the intricate relationship between life and technology. Our assertion here is that Canguilhem rejects the idea that organisms are machines in a mechanistic sense, yet he does not dismiss the possibility of non-mechanistic organic machines, as is implied in his scattered use of the term "general organology". His analyses of the relationship between life and milieu ultimately prompt contemplation of these very machine monsters. In our reading, this line of thought culminates in an ontological understanding that can be derived from Canguilhem's epistemology. It involves contemplating nature as a vast mega-machine that can only be defined in relation to its milieu, equilibriums, norms, and pathological states. In essence, nature as an organic machine that accommodates and even embraces anomalies and aberrations, and monstrosities.[8] Now, let us explore how this concept is further developed in "Machine and Organism".

In this lecture, Canguilhem presents a counterargument to Descartes' view, asserting that living beings cannot be reduced to mere machines.

[8] Catherine Malabou in her short section of Canguilhem in *Morphing Intelligence* ascribes "the idea of ontological monstrosity" to Canguilhem (Malabou 2019, 43).

Indeed, as mentioned, Canguilhem's stance is opposed to Descartes' mechanistic perspective. However, he endeavors to propose a non-mechanistic, or rather organologic, understanding of technology and machines based on Descartes' idea of living machines, which Canguilhem also discussed in *Descartes de la technique*.[9] This necessitates acknowledging the fact that technicity is a product of nature, and furthermore, recognizing nature itself as a wondrous machinery. In this sense, according to Canguilhem, technique can be regarded as a "universal biological phenomenon" (96).

Canguilhem cites Descartes, who states that the rules governing machines are the same rules governing nature (92). On the contrary, the rules governing manmade machines are actually the rules established by their designer, namely, humans. For Descartes, the rules of nature are in fact the principles of physics. Canguilhem expands upon this Cartesian dualism by introducing a third realm, which he identifies as the realm of life.[10] Organisms, in contrast to manmade machines, do not adhere to the laws of physics. To support this idea, Canguilhem references Kant's *Critique of the Power of Judgment*, specifically to illustrate that human creative activity is not completely exempt from the rules of organic nature. In paragraph 75 of "Critique of the Teleological Power of Judgment", Kant draws a distinction between intentional technique employed by humans and unintentional technique exhibited by living organisms. The manifestation of unintentional technique in human life is exemplified through artistic activities, which Kant argues are distinct from the rational activities of humans and their capacity for knowledge. "Properly speaking, rationalization is a mechanization of the organism" (96). The ultimate conclusion drawn is the fundamental distinction between organisms and mechanisms, highlighting the inexplicability of the former through rational knowledge contrasted with the explicability of the latter. Then, Canguilhem claims, "We therefore find that the works of ethnographers shed more (though still weak) light on the construction

---

[9] Canguilhem uses sometimes the term "machinist" instead of machine in order to refer to organic machines; machine is mechanistic, machinist is organic: see for example p. 111.

[10] Although Canguilhem's references to modern physics indicates his revised Cartesianism which replaces Descartes mechanistic physics, not only with biological grounds, but also with the probabilistic physics.

of machines than those of engineers" (93). The ethnographer investigates the emergence of technique and its historical origins, whereas the engineer analyzes the usage of a machine based on its design. The ethnographers "have been attentive to the relationship between the production of the first tools, the first devices [dispositifs] for acting on nature, and organic activity itself" (93). Canguilhem refers, among others, to André Leroi-Gourhan, and his *Milieu et Techniques* which in its final chapters deals with the relationship between biology and technology, and to Georges Friedmann's *Problemes humains du machinisme industriel* which focuses more on the relations between machines and human organism. Within these ethnographic contexts, Canguilhem endeavors to portray technique as a specific behavior exhibited by living beings. Ethnography demonstrates how the technical behavior of living beings contributes to the emergence of sapience. Knowledge is an inherent aspect of life, and it is through this process that nature generates culture, encompassing various forms such as human culture.

Consequently, the ethnographic approach reveals an irrational technicity that distances it from being a meticulously designed product of personal intention, as Canguilhem declares, "The rationalization of techniques makes one forget the irrational origin of machines" (95). Therefore, the mechanical perspective is merely an anthropomorphic reductionist viewpoint within the realm of the organic. Mistakenly, it confuses the resultant form with the original force. Moreover, it is within the historical and social existence of the organic that technical machineries are generated through its own metabolic processes. An individual organism inherently differs from an individual manmade machine. However, the histories of life and technique are intricately interconnected.

Can a Machine Get Sick?

Although organisms and mechanisms belong to distinct categories, organisms and machines can be included within the same category. This is the ultimate meaning of what Canguilhem calls "general organology". By "machine", we do not mean a single device used as an instrument, but rather machinery as a whole. This is why, it is important to insist that

Canguilhem strives to approach a univocal ontology. This is evident in his suggestion that humans operate at the same level as other organisms, which is indeed a Cartesianism at the level of organology. Canguilhem declares this by comparing humans and animals:

> Doubtless, the animal cannot resolve all the problems we present to it, but this is because these problems are ours and not its own. Could man make a nest better than a bird, a web better than a spider? (Canguilhem 2008, xviii)

If we utilize human problems and the capacity to confront those problems as the criteria for assessment, we would establish a hierarchy between humans and animal species that encounter similar problems and are comparable to humans in their abilities.[11] A thorough philosophical or scientific study would encompass the inherent contextuality of problems, recognizing their specific and localized nature. Analogism always poses a peril, in this regard. We are superior to animals only in terms of what we do, but not in terms of what they do. This becomes especially apparent when we shift our focus from individual organisms to colonies and societies.

The next step is to expand this view to technique as natural product. The thought of monstrous machine is pivotal here. An organism can get sick. A machine can be defect, or not working properly. An organism can die, while a machine can become dysfunctional. But this is only applicable to manmade functional machines, and the question arises, what about hybrid machines, or the machinic structures that nature fabricates? Are there living machines? Canguilhem's quantitative approach turns the difference between machines and organisms into a difference of degree. The problematization of the difference between normal and pathological, along with the challenges that he introduces regarding the difference between life sciences and physicochemical sciences, leads us to

[11] This is exemplified by John Searle's approach to the issue of animal consciousness in his 1994 article "Animal Minds", which renders his analysis weak and superficial. Searle fails to examine the problem of animal consciousness from the perspective of animal-specific challenges. Similarly, when discussing the ability to communicate through language, he overlooks the various non-linguistic communication methods employed by different animal species (Searle 1994, 206–219).

problematizing the sharp distinction between organism and machine. This problem becomes evident particularly with relation to disease and malfunctioning and the way Canguilhem understands both in terms of the role of the milieu. Canguilhem in "Monstrosity and the Monstrous" claims that "There are no mineral monsters. There are no mechanical monsters" (Canguilhem 2008, 135). For him, a monster is obviously a living being. It is of course an abnormal living being, a counter-organism. The monstrosity of the monster issues from the nonviable, the exterior, of the living being. This takes him to extend the quality of monstrosity to natural sciences, and therefore, to nature itself. The conclusion would be considering nature a monstrous living being, exterior to living organisms, but itself somehow living: "We have said that life is poor in monsters, while the fantastic is a world" (145). "Organisms are incapable of structural eccentricities except during a short moment at the beginning of their development". But the totality of nature, or a world, as Canguilhem states, can be more fantastic than an organism due to its long-lasting monstrosity. In this regard, the machines of nature, and the totality of human technicity, are living a non-vital life. The principal idea here is considering the totality of technique as the monstrous product of nature. Seeing from the viewpoint of nature, is not human culture and its technicity the most monstrous phenomenon in the world? Is it not the greatest sickness of nature? Is not nature itself more an "anticosmos", "a chaos of exceptions without laws", than similar to an organism? Isn't the monstrous agency, which constitutes the decision-making algorithm of a monster, the seed of all agencies and intentionalities, including the social-political one? This is Canguilhem's third way out of the duality of mechanism and organism.

References

Bachelard, Gaston. *The Formation of the Scientific Mind: A Contribution to a Psychoanalysis of Objective Knowledge*. Trans. Mary McAllester Jones. Manchester: Clinamen Press, 2002.

Bennett, Jane. *Vibrant Matter: A Political Ecology of Things*. Duke University Press: Durham and London, 2010.

Canguilhem, Georges. *Knowledge of Life*. Translated by Stefanos Geroulanos and Daniela Ginsburg. Fordham University Press, 2008.

Canguilhem, Georges. *The Normal and the Pathological*. Translated by Carolyn R. Fawcett. Zone Books, 1991.

Canguilhem, Georges. *A Vital Rationalist: Selected Writings from Georges Canguilhem*. Ed. Francois Delaporte. Translated by Arthur Goldhammer. Zone Books, 1994.

Canguilhem, Georges. *La formation du concept de re´flexe aux XVIIe et XVIII sie`cles* (2nd ed.). Presses Universitaires de France, 1977.

Canguilhem, Georges. *Ideology and Rationality in the History of the Life Sciences*. Trans: Goldhammer A. MIT Press, 1988.

Descartes, René. *The World and other Writings*. Translated by Stephen Gaukroger. Cambridge: Cambridge University Press, 2004.

Descartes, René. *Principles of Philosophy*. Translated by V. R. Miller and R. P. Miller. D. Reidel Publishing Company, 1982.

During, Elie. "'A history of problems': Bergson and the French epistemological tradition". *Journal of the British Society for Phenomenology, 35*(1), 4–23 (2014).

Eyers, Tom. *Post-Rationalism: Psychoanalysis, Epistemology, and Marxism in Post-War France*. Bloomsbury, 2013.

Elden, Stuart. *Canguilhem*. Polity, 2019.

Fiant, Océane. "Canguilhem and the Machine Metaphor in Life Sciences: History of Science and Philosophy of Biology at the Service of Sciences". *Transversal: International Journal for the Historiography of Science*, 4 (2018) 149–162.

Hacking, Ian. "Canguilhem amid the cyborgs". *Economy and Society*. 27(2–3), 202–216, (1998).

Hui, Yuk. *Recursivity and Contingency*. Rowman & Littlefield, 2019.

Malabou, Catherine. *Morphing Intelligence: From IQ Measurement to Artificial Brains*. Translated by Carolyne Shread. Columbia University Press. 2019.

Nicholson, Daniel J. "Book Review: Life Sciences". *Annals of Science* ISSN 0003-3790, 2012.

Parsa, Mehdi. "Transcendental Extinction: A Philosophical Response to the Anthropocene", in *Athena: Philosophical Studies*, No. 14, 2019.

Searle, John. "Animal Minds", *Midwest Studies in Philosophy* 19 (1):206–219 (1994).

Schmidgen, Henning. "The life of concepts: Georges Canguilhem and the history of science". *History and Philosophy of the Life Sciences*, Volume 36, pages 232–253, (2014).

Talcott, Samuel. *Georges Canguilhem and the Problem of Error*. Palgrave Macmillan, 2019.

Wolfe, Charles and Gandolfi, Giulia. "The Case of Life in the Historiography of Modern Science: Canguilhem's 'Biophilosophy'". In M. L. Condé, M. Salomon (eds.), *Handbook for the Historiography of Science*, Springer Nature, 2023.

Vázquez García, Francisco and Toscano López, Daniel. "Georges Canguilhem and Transhumanism: A Confrontation". *Rundschau*, 119 (6), pp. 79–103, (2021).

3

Gilbert Simondon and the Transindividual Machine

The works of George Canguilhem and Gilbert Simondon are among the most significant contributions to understanding the relationship between life and the development of technology. In this chapter, we focus on Simondon's work, arguing that the philosophical elaboration of this problem is found not only in his first monograph, *The Mode of Existence of Technical Objects* (Méot, 1958), but also in *Individuation in Light of Notions of Form and Information* (1964 and 1989). We argue that the concept of the "transindividual" signifies the process through which life gives rise to technology. In this course, following Simondon's *The Mode of Existence of Technical Objects*, technology is defined in terms of network and the related concept of the "individualization of technical objects in an ensemble" (*Mode*, 65). Through a generalized notion of the network and the related concept of information, echoed in *Individuation* through the idea of the "transindividual", Simondon addresses the genesis of the technical object. In his view, the technical object is not merely an artifact or tool, but rather a machine. Therefore, rather than focusing on its manufacture, instrumental meaning, and use-directed individuation, one should discuss its (natural and contingent) genesis.

Thus, Simondon presents a historical framework of nature in which life generates technics, and in turn, technics gives rise to human beings.

M. Parsa, *Machinic Ontology*, https://doi.org/10.1007/978-3-031-99561-3_3

Regarding the first move, Jean-Hugues Barthélémy states, "Simondon wanted to think the living being such that it engenders technics" (Barthélémy 2015, 18). Concerning the second move, Jason Read emphasizes that the human being is a product of technology, not the other way around: "Technology is the source of both our idea of the human individual, the toolmaking animal standing above all of nature, of its undoing in the great machine, real and symbolic, that reduces human beings to mere appendages, a cog in the machine" (Read 2016, 107). In this context, Simondon's philosophy can be regarded as a machinic ontology, encompassing various levels of machines—physical, biological, psychosocial, and technical—where manmade tools are merely one of its superficial components. As Alberto Toscano observes, the essence of this machinic ontology lies in presenting "relation itself as an ontogenetic factor" (Toscano 2006, 137). Life can generate technology because the entire realm of ontogenesis is founded on a relationality that imbues the physical, biological, and other domains of individuation with a machinic spirit.

Philosophically, this issue can be explored through the following question—which may be the primary motivation behind *Individuation in Light of the Notions of Form and Information*: What is the mode of being in itself that enables it to generate technical objects? We can pose this question because Simondon's philosophy is monistic, offering a singular concept—individuation or ontogenesis—to explain all phases and aspects of being, including the physical, biological, psychosocial, and technological. This is called the "transductive unity" of ontogenesis (*Individuation*, 12; *Mode*, 210). From the beginning of *Mode* and throughout *Individuation*, Simondon consistently emphasizes this unity by eliminating the assumed divide between man and machine, man and nature, and machine and nature. This is evident in Simondon's discussion of the incorporation of machines into culture (*Mode*, 19), his concept of "open machines" (17), and his assertion that humans exist "at the same level as machines" (125). On this basis, the development of technology from life is comparable to the emergence of the biological from the physical order. All these developments are made possible by what Simondon attributes to the physical domain: metastability. This necessitates an entanglement of all domains of individuation, such that what Simondon identifies as the transindividual—the social and cultural domain—is already present in the development of the biological. Thus, the main argument of this chapter is as follows: if life is capable of generating technology and the physical is capable of generating life, then being must have a technological foundation. In Simondon's thought, this technological foundation takes the form of an ontological relationalism, where a relation is understood in terms of metastability.

What is the technological ground of being? The answer lies in the transductive unity of being, elaborated through the unity of the domains of science, the concept of pre-individual being in terms of tensions and metastability, and how this pre-individual being underpins the transductive unity. In *Mode*, the margin of indeterminacy—the inherent contingency—of open machines is what renders them natural entities. The inner tensions and metastability are described as "the absolute beginning" of a machine, explained in terms of the "asymmetrical conductance" between the anode and cathode in a diode (*Mode*, 44–45). The genesis of a technical object does not lie in its planned manufacture or engineered formation. Rather, it reflects technics in its evolution—in simple terms, in the history of technicity, which begins with physical machines driven by metastability. This indicates a logic of contingency that governs the entirety of technical evolution. The contingent evolution, or history, progresses through "self-adaptation" and "inner resonance" (26). Using the example of car manufacturing, Simondon distinguishes between development based on user requirements and demands, and an inner historical technical evolution. In this way, he delineates a natural automacy that characterizes technical evolution. As Jason Read states, "As the machine develops, its history seems to be determined less by the inventions of individuals than by its own particular perfection and development; it seems to be less the creation of individuals than its own process of individuation" (Read, 108). This independence, this self-development, brings the notion of machine closer to something that is identical to any material structure evolving through natural history.

The issue with *Mode* is that it focuses too heavily on the relationship between machine and man. The so-called natural historical aspect is nearly absent from it. Indeed, the ontological problem we are exploring (the relationship between life—or nature—and technology) must be sought more in *Individuation* than in *Mode*. Thus, our goal is to seek to understand the genesis of machines through the pages of *Individuation*. However, we must keep in mind that the problem is posed in *Mode*, where the technological as the transindividual—whose place in *Individuation* is at the final stage of collective individuation—appears as the beginning, the very pre-individual.

The Pre-Individual Machine

The pre-individual state is not yet individuated. This self-evident fact is the starting point of Simondon's *Individuation in Light of Notions of Form and Information*. In the opening pages, Simondon, in a Heideggerian fashion, describes the being of the individual, or the individuated being, as relative (3). Simondon's Heideggerian question is: What is absolute being? How can we describe a state of being that is not yet individuated? How can philosophy approach the pre-individual being? Simondon's answer to these questions lies in the direct connection he establishes between the absolute pre-individual and transindividuality, thereby constructing a transductive unity. In other words, there is a distinction between relation and the relative. The pre-individual being is the absolute relation that gives rise to the relative. His concept of the absolute relation is one of his most significant philosophical contributions, which drew the attention of contemporary philosophers, including Gilles Deleuze in *Difference and Repetition*. To be generative, the pre-individual relation must consist solely of metastability and tension. Unlike the classical notion of relation as an accident occurring between things, here we are concerned with real relations that generate things. In this manner, Deleuze in the fifth chapter of *Difference and Repetition* writes, "Gilbert Simondon has shown recently that individuation presupposes a prior metastable state – in other words, the existence of a 'disparateness' such as at least two orders of magnitude or two scales of heterogeneous reality between which potentials are distributed" (Deleuze 2014, 322).

Being as such as pre-individual being is an infinite amount of pure potential energy, comparable to thermodynamic chaos.[1] In this way, the pre-individual can be defined based on potentialized "disparateness" that leads to metastability. The main characteristic of thermodynamic chaos is that, although it does not contain individualized things, it cannot be described as a single, unified homogeneous entity. This description applies not only to the pre-individual being but also to being as a whole.

According to Simondon's well-known example to explain metastability, pre-individual being is like a hyper-saturated solution. "Reality is initially, in itself, similar to the supersaturated solution (and even more so in the pre-individual regime)" (*Individuation*, 5). It has the highest degree of entropy. It enjoys a phase-change when sucrose crystals, for example, are formed from the solution. This phase-change is the first step of individuation and leads to a solution which is, in one of the possible directions, less saturated. In the language of thermodynamics, its entropy decreases (there are of course dissolutions in which the entropy of the system increases, for example when each molecule in the solution is transformed into several ions, or when a solid is dissolved in a liquid. But the dissolution of a hyper-saturated solution is the other way around regarding the potential energy of the system). The pre-individual being is then, by definition, a solution whose entropy is at a maximum and cannot be further increased. It is infinite because it potentially contains all possible individuals or all possible ways of individuation. In order for this, the pre-individual contains tensions and disparities that lead to the separation of phases and the emergence of a plurality of individuated beings. The tensions and disparities are prior to the formation of individuated things. As Toscano writes, "to relate … simply *is* to individuate" (Toscano 2006, 139). There are then tensions and disparities without there being different things, relations without things to be related. There are rather different fluxes of potential energy in thermodynamic chaos. Hence, there is no pre-individual that generates only one individual, which means that there is no dissolution where the entropy of the system remains unchanged. Chaos contains no form and no structure, but rather has the infinite potential for formation and structuration.

---

[1] Similarly, J. L. Nancy claims, "… the real 'in itself' is chaos, a sort of effectivity without effectuation" (Nancy 1996, 110).

The pre-individual being, or thermodynamic chaos, which is not a single unified entity, evolves continuously due to its inherent metastability. Therefore, individuation leads to a plurality of not only diverse beings but, more importantly for Simondon, a plurality of phases: "Ontological monism must be replaced with a pluralism of phases" (*Individuation*, 358). The problem with classical ontologies is that they are based on the identity of being, a doctrine that ignores this evolving pluralism of phases: "identity, an impoverished relation, is the only relation of the being to itself that can be conceived according to a doctrine that considers the being as having a single phase; in the theory of polyphasic being, identity is replaced with internal resonance, which, in certain cases, becomes signification and permits an amplifying activity" (*Individuation*, 359). Therefore, what is said about pre-individual being is also true about being in each and every of its phases. The pre-individual being persists throughout any phase of individuation. No generation of individuals is full and complete. Every individual is by nature incomplete. It is only the pre-individual that is complete, but its completeness lies in being infinitely full of potentials:

> it can be supposed that individuation does not fully exhaust pre-individual reality and that a regime of metastability is not merely sustained by the individual but carried by it, such that the constituted individual transports along with it a certain associated charge of pre-individual reality that is animated by all the potentials which characterize it. (8)[2]

This is why the notion of metastability, borrowed from the physical paradigm, is applicable to other phases of individuation. Important for our purpose is the mutual exchange between physical-biological and

[2] This is a very frequent theme in *Individuation in Light of Notions of Form and Information* and this fact indicates the centrality of this thought in Simondon's work. He repeats again and again that individuation "cannot exhaust the possibilities of pre-individual being" (*Individuation*, 348); "the individuation of beings does not completely exhaust the potentials of organization" (357); "individuation does not exhaust from the start the potential resources of the being in an initial operation of individuation: the first pre-individual state of the being can continue to exist in association with the result of an initial individuation" (368), etc. See also Jason Read: "the pre-individual is not so much prior, understood as something that comes before individuation, than it is that which exists alongside individuation as a different phase" (Read, 109); or, "The persistence of the pre-individual in the individual" (Read, 109).

technical individuation—specifically, how the technical object inherits the contingency of its functioning from physical-biological metastability, while physical and biological entities, in turn, can be regarded as small natural machines. However, the larger machines are the phases themselves. Individuation is a concept relevant to the constitution of each phase as a meta-consistent system, defining its own logic and ontology. In this sense, one can speak of a physico-biological machine or a psycho-social machine (or system) alongside the technical machine.

A Contingent Positivism

As previously stated, *Individuation in Light of Notions of Form and Information* was published in two parts, the first part in 1964 with the title *L'individu et sa génèse physico biologique* (*Individuation and its physical-biological genesis*) and the second part in 1989 under the title *L'individuation psychique et collective* (*Psychic and collective individuation*). The title of the first book indicates that, in the case of physical-biological individuation, we are dealing with one and the same process of individuation.[3] This is indeed the source of the accusation of positivism and objectivism that some of Simondon's contemporaries leveled against him.[4] Not only is the biological a continuation of physical individuation, but also the psychical-collective represents only a phase change in the same process of individuation. This is to say that, for Simondon, physics is not only the starting point, but also the paradigm for the entire philosophy of individuation. This is possible through the way the metastability as a physical notion is carried along other phases of individuation. Muriel Combes, in his commentary on Simondon, declares that physics plays the role of a paradigm for the entire realm of ontogenesis, including vital, psychic and collective individuation (Combes 2013, 12–14). What the accusation of positivism or objectivism (or what is sometimes referred to by Simondon as "materialism"; see *Individuation*, 169) neglects is that Simondon does not

[3] Most of commentators agree that the same holds for the second volume although an "and" here takes the place of hyphen

[4] See Combes, pp. 53 and 90.

reduce vital, psychic, and collective individuation to merely physical individuation. Instead, in the same manner as Canguilhem, he rethinks and reformulates the physical paradigm so that the realms of the vital, the psychic, and the collective, with all their complexity, can be derived from it.[5] The physical paradigm must be imbued with a profound complexity that centers on the idea of contingency. The first step in achieving this is to rid physics of the notions of unity and identity. Only then can physics serve as the paradigm for ontogenesis.

An interesting example that Simondon provides in his Introduction to *Individuation in the light of the notions of Form and Information* refers particularly to the theory of wave-corpuscle complementarity in modern physics. This is used to demonstrate that the move toward reality entails distancing from unity and identity. In its historical evolution, it was wave mechanics that initiated one of the first attacks in physics on the presupposition of self-identical particles which are already individuated. Simondon uses this to criticize the notions of unity and identity, implying the metastability of the real: "Complementarity would then be the epistemological reverberation of the initial and original metastability of the real. Because they are theories of identity, neither mechanism nor energeticism fully account for reality" (*Individuation*, 6). Reality is not in the form of identity, it is not a sum of identities. Furthermore, it is not endowed with unity; it is not unified. The principle of contradiction or excluded middle is not a principle of reality in itself, but rather a principle

[5] "custom in fact prompts us to think that living beings cannot result from physical beings since they are superior to the latter due to their organization. Nevertheless, this very attitude is a consequence of an initial postulate according to which inert nature cannot contain a high level of organization. If, on the contrary, we posited right from the beginning that the physical world is already highly organized, we wouldn't be capable of committing this basic error that results from a devaluation of an inert matter" (*Individuation*, 169). Or more clearly in Conclusion, "we have attempted to extract a paradigm from the physical sciences by thinking that it can be transposed into the domain of the living individual: the study of this physical domain is meant not only to form notions, but also to serve basically as the study of a first domain within which an operation of individuation can exist; since we suppose that there are various degrees of individuation, we have utilized the physical paradigm without reducing the vital to the physical, because the transposition of the schema is accompanied by a composition of the physical itself. We do not mean to say that physical individuation is what produces vital individuation: we simply mean to say that reality has not clarified and developed all the possible steps of the operation in the physical system of individuation, and that a vital individuation still remains available within the physically individuated real" (*Individuation*, 360).

of reality after it has been individuated. If this is the case, concepts of wave or particle, like any other concept, are inadequate for describing reality as such: "concepts are adequate only to individuated reality and not to pre-individual reality" (*Individuation*, 6). The theory of supplementary in physics then indicates an irreducible gap between the conceptual image of individuated being and the non-conceptual nature of being as such. In this way, modern physics, while striving for a non-conceptual description of reality, offers a paradigm for the entire realm of individuation, which encompasses reality itself, including the vital, the psychic, and the collective. Simondon supposes that "reality is initially, in itself, similar to the supersaturated solution (and even more so in the pre-individual regime), more than unity and more than identity" (*Individuation*, 5). Or elsewhere, "a being in a state of pre-individual tension, i.e. in a being which is more than unity and more than identity" (14). Modern physics provides a paradigm that takes philosophy beyond unity and identity. It provides the paradigm of contingency, which can be defined as the synthesis of probability and potentiality. In this manner, the concept of chaos emerges from physics to become the metaphysical ground of ontogenesis.

In the first section of *Physical Individuation*, Simondon aims to move away from classical notions of form and matter towards a theory of individuation and information, using the example of clay and mold to explore the actualization of potential. In his critique, he challenges the classical concept of hylomorphism that assumes the existence of distinct domains of homogeneous matter and fixed form. Hence, he concludes, "individuation cannot have its principle in matter or in form; neither form nor matter are enough for form – taking … what is individuating is the energetic system" (*Individuation*, 32). According to his perspective, information represents an intermediate dimension that exists in a pre-individual state, preceding the formation of forms and homogeneous matters. This realm is characterized by informed matter and potential energy. Simondon's "technological paradigm" asserts that physical individuation is the underlying process by which potential energy is actualized into a physical object. This process involves a singular, irreversible operation that results in the completion of an object. However, Simondon argues that the individuation of living beings differs from this technological paradigm:

> [under the technological paradigm] The individuation is completed in a single stroke; the individuated being is never more perfectly individuated than when it leaves the hands of the craftsman. Thus, there is a certain exteriority of the operation of individuation relative to its result. Conversely, in the living being the individuation is not produced by a single operation that is limited in time; the living being is to itself partially its own principle of individuation; it continues its individuation, and, instead of merely being a result that progressively degrades, the result of an initial operation of individuation becomes the principle of a further individuation. (*Individuation*, 33)

While Simondon seeks to distance himself from Aristotelian hylomorphism, his use of the technological paradigm—which draws on an Aristotelian example of a craftsman molding matter—hinders him from achieving an immediate synthesis of physical and biological individuation. Simondon's argument that the individuation of a break from the mold is a singular operation, with the break remaining separate from the mold, is valid. However, the question remains whether this paradigm is suitable for explaining the production of non-biological objects in nature. When reading "The Physical-Biological Individuation," it is crucial to note that the technological paradigm that Simondon employs to differentiate between physical and biological individuation is entirely separate from the physical paradigm that serves as the basis for the entire domain of individuation. The latter is introduced in the Introduction and draws on modern physics.

The example of mold and clay could be used to understand the nature of biological individuation (rather than physical individuation). In this regard, the biological object contains an internal resonance that requires a paradigm different from technological production, but not different from the physical paradigm. This means that we are dealing with organic individuation. In this case, the metabolism between the individuating and individuated (unlike mold and clay) is called Allagmatic. Later, Simondon concludes that the Allagmatic metabolism, which is derived from biological individuation (and is only temporarily present in the technological paradigm), is a better paradigm for the entire domain of physical-biological individuation. "The essence of physical individuation, which is identical to biological individuation, is Allagmatic. It possesses a kind of organization that distinguishes it from the technological paradigm. This is the source of probabilistic developments in modern physics ... analogically, it would be necessary to consider the physical individual as a biological society, and the latter alone would be the image of a (albeit very simple) totality" (*Individuation*, 168). Therefore, although Canguilhem develops a philosophy of science based on life sciences, and Simondon introduces physics as the paradigm of all sciences, Simondon's understanding of physics based on the Allagmatic principle aligns him on the same camp with Canguilhem. Following Canguilhem, Simondon recognizes the crucial role that developments in life sciences play, which leads him to propose the central Allagmatic principle that affects his understanding of physics.

Nevertheless, it can be argued that the response to the accusation of objectivism and physicalism is not necessarily a reduction to alleged vitalism. It is not the case that Simondon applies what he derives from organic individuation to physical production. There is no vitalism at work here. Simondon describes the pre-individual force by declaring that "This force is not vital; it is pre-vital; life is a specification, a first solution, complete in itself by leaving a residue outside its system", or, "This pre-vital charge should not be called *élan vital*, since it is not exactly in continuity with vital individuation, although it extends life, which is a first individuation" (*Individuation*, 340). Furthermore, Simondon refers to modern physics to demonstrate the contingency of the primordial chaos, and he notes that physics reached these (modern) developments through its quasi-immanent evolution. In this way, the physical individuation, which Simondon uses the example of crystallization to illustrate, serves as a paradigm for biological, psychic, and collective individuation, a paradigm that is not exhausted in the development of the physical individuation into organic individuation, etc. Therefore, he applies the physical concepts of metastability and "syncrystallization" to the domains of psychic and collective individuation (*Individuation*, 271 and 333).

Logic of Ontogenesis

Simondon's emphasis on the paradigmatic significance of physics underlies a focus on ontology, or more precisely, entails an ontological shift that opposes the logicism of post-Kantian philosophy. The focus is on the nature of being and its genesis. For Simondon, following the tradition of Cavaillès and Canguilhem, knowledge is a byproduct of the genesis of being: "to know how being can be thought, one must know how it individuates, for this individuation is the support of the validity of any logical operation that must conform to it" (*Individuation*, 362). To support this view, Simondon defines knowledge as tropism or taxic orientation (10–11). Knowledge regulates the individuated entity in relation with the fluxes of individuation within and without it. According to Simondon, this involves a concreteness in knowledge that is based on the locality of the entity within the milieu or the world. The object of knowledge is "the source of the initial gradients" which "organizes them according to a world" (10). The same holds true for the internal emotional states of the individuated being, which involve changes in mood that are not possible without an internal relation to the pre-individual being. In this manner, Simondon describes knowledge as a partial side-effect of individuation which, in itself, has no central role for the individual. Or in Combes' word, "knowledge *exists in the same mode as the beings that it links together; considered from the point of view of that which constitutes their reality*" (Combes 17–18). The knowledge of something exists at the same level as that thing, and the thing in question should be considered as relational. Simondon's view here is in line with Canguilhem's description of knowledge in relation to life.[6] Similar to Canguilhem's identification of knowledge and life, Simondon's allagmatic principle, which presents the contingency of biological individuation as the characteristic of the physi-

[6] Bardin refers to the ontological nature of Simondon's thought as his debt to Canguilhem. In fact, he reads Simondon as an extension of Canguilhem beyond the domain of living organisms and into the ontological realm: "[Simondon] not only refers normativity to 'living systems' but to matter itself (not 'inert' anymore), thus building an ontological paradigm which can cover other fields of research. In Simondon, Canguilhem's heritage extends to all the regimes of individuation, from the physical to the social: the different regimes of individuation always cross in metastable relationships since the different normativities produce systemic effects" (Bardin 2015, 77).

cal paradigm, places the object and subject of knowledge at the same level of individuation. Therefore, it is not only thought that is derived or abstracted, but individuated being is also abstracted (although, abstract here does not mean unreal). Hence, linking knowledge to being would be the same as linking individuated being to its ontogenetic source.[7]

Hence, relation is constitutive of its terms, and logic, as the science of relations, attains an ontological status. According to Simondon, any relation or effect has a rank of being, or more precisely, the highest rank of being. The logic of being is not a subjective instrument. Although this view is prevalent in Simondon's thought, it is not entirely compatible with the parallelism or analogy he draws between the individuation of thought and that of being. At this point, we would like to add a nuance to his chain of thought and discuss the entanglement of these two individuations, rather than their parallelism, as the source of logic. In this way, we would have a complex structure in which the genesis of thought, which, according to Simondon, is parallel and similar to the genesis of being, is entangled with the genesis of being. Simondon does not appear to provide an account of this entanglement in his *Individuation in Light of Notions of Form and Information* (an account which is attempted in *Logic of Sense* by Deleuze).[8] The important point that Simondon mentions in this regard is that the structuration of thought, which is logic, is the same as the function of its individuation.

Hence, Simondon distinguishes between the logic of individuals and the logic of individuation. And only the latter has the status of the logic of being. In ontogenesis, logic and ontology are one and the same. This highlights the main failure of classical logic with its principles of excluded middle and identity:

> This kind of method supposes an ontological type of postulate: at the level of being grasped before any individuation, the principle of the excluded middle and the principle of identity are no longer applicable; these principles only apply to the already individuated being, and they define an impover-

[7] "the relation between thought and the real becomes a relation between two organized reals that can be analogically linked by their internal structure" (*Individuation*, 170).

[8] See Deleuze's *Logic of Sense*, Fourth Series of Dualities and Sixth Series on Serialization, pp. 25–29 and 39–44.

> ished being that is separated into milieu and individual; consequently, they do not apply to the being's whole, to the ensemble formed later on by the milieu and the individual, but merely to what of pre-individual being has become individual. In this sense, classical logic cannot be used to think individuation, for it forces us to think the operation of individuation with concepts and rapports among concepts that merely apply to the results of the operation of individuation considered partially. (*Individuation*, 13)

Classical logic fails because it is based on a false and limited ontology, namely that of substances. Being in its completeness, which is to say, considered together with its process of individuation, contains its logic within itself. Its logic is not composed of abstract concepts, but concrete effects, taxic orientations, and information.

The universality of the fact that relations are constitutive of their terms gives rise to what Simondon refers to as the concept of the "transductive unity" of being (12). This concept can only be comprehended through the lens of a theory of individuation and the phase-transitions that such a theory suggests. In essence, it implies that the sole genuine connection lies in the process of genesis, and the ultimate unity arises from primary genetic relationships. The goal of this transductive unity of being is to contemplate the entirety of being, encompassing both the individual entities and their individuation, both the state of being and the process of becoming, bridging the realms of ontology and logic, all within a single comprehensive perspective called ontogenesis. The transductive unity of being is based on the capacity of being to undergo a "phase-shift in relation to itself"—the capability to extend beyond its own boundaries (12). It represents a genetic unity, the unity of a flux, or more precisely, the unity encompassing all flows and becomings. This singular unity, founded upon the concept of transduction, is the unity liberated from the confines of identity.

Logic diverges from ontology in its abstraction from the actual system or the real fluxes of individuation. Logic becomes viable when an individual is isolated from its milieu, enabling it to analyze and resolve issues connected to that milieu. In a state of pure chaos, there would exist no demarcation between logic and ontology. This distinction only emerges within the framework of chaos-cosmos. Now, transductive logic "extracts the resolving structure from the very tensions of this domain, just as the supersaturated solution crystallizes due to its own potentials and according to the chemical species that it holds, not with the contribution of some foreign form" (15). This extraction provides the primary source of logical abstraction.

Although it is abstract, it is not formal. It is based on an abstraction that maintains connections with concrete flows.[9] This is why it is logic without generality:

> It is also not comparable to induction, for induction truly conserves the characteristics of the terms of reality included in the studied domain, drawing the structures of the analysis from these terms themselves, but it only conserves what is positive in these terms, i.e. what is common to all terms, thereby eliminating what is singular from them; on the contrary, transduction is a discovery of dimensions whose system makes the dimensions of each of the terms communicate, such that the complete reality of each of the terms of the domain can become organized into newly discovered structures without loss or reduction. (15)

Therefore, Simondon distinguishes transduction from both deduction and induction. In both deduction and induction, there is a reduction of information concerning the singularity of terms. Deduction disregards this singularity right from the outset by commencing with general terms and principles that transcend the system. Meanwhile, induction does not depart more from this assumption of generality, as it preserves what is shared among all terms. Both approaches lead to "an impoverished real without potentials, and consequently one that is incapable of individuating"

[9] Gaston Bachelard in *The Formation of Scientific Mind* studies the characteristics of the so-called scientific abstraction (Bachelard 2002, 11). In his phenomenological view, scientific thought is abstract, in the sense that it distances itself from common sensical view of reality. It is in this sense that science is profoundly revolutionary and society treated great scientists in the same manner that it treats political rebels (Galileo). The world that quantum mechanics or theory of relativity pictures is so abstract that provides the perfect realm for fictional imagination (the emergence of the so-called science-fiction as a genre). Bachelard introduced this aspect as the essence of scientific practice. In this regard, we can distinguish between what we prefer to call immanent abstraction vs transcendent abstraction: Gaston Bachelard is the thinker of immanent abstraction. Abstract universals, such as platonic ideas, are only the products of a machinic immanent abstraction, and therefore are only illusionary and not scientifically abstract.

(16). In this sense, both approaches are formalistic and hylomorphic. They begin and remain with the constituted individuals. Transduction is the only way to access the informative nature of being, its singularity, as is evident in its individuating potentials.[10] This is why, according to Simondon, "*The notion of form must be replaced with that of information*" (16).

The informativeness of transduction implies that we are perpetually engaged with a genetic multiplicity (comprising various flows), harboring within it an information framework that exists independently of and precedes the formation of distinct entities. In other words, what carries meaning and significance is detached from and precedes the individuals and their associated concepts. The process of individuation encompasses an informational phase change. For Simondon, individuation is based on a unified, but not self-identical, realm—its transductive unity—which involves a division between becomings or flows, but not genres and entities, which only appear as outcomes. Therefore, the demarcation between domains boils down to no more than variations in the relative rhythms of individuation. As Combes declares, "What divides being into domains is ultimately nothing other than *the rhythm of becoming*" (Combes 23); and this idea makes her to claim that "The difference between plants and animals is explained in a manner similar to the difference between the physical and the vital" (Combes, 22). This is nothing but the transductive unity of being at its ontological core.

Although being is transductively unified, Simondon spends a large part of his "Physical-Biological Individuation" to explain how biological individuation is different from physical individuation. In our understanding, the difference is comparable to the aforementioned differentiation between logic and ontology, which only arises in the individuated chaos-cosmos (rather than in pure chaos). At this point, it is noteworthy to observe that, according to Simondon's perspective, an organism serves as the "theater of individuation" in the sense that it exhibits a transductive unity (a chaos-cosmos) in which individuation occurs (*Individuation*, 7).

[10] Deleuze emphasizes this point: "Such a pre-individual state nevertheless does not lack singularities: the distinctive or singular points are defined by the existence and distribution of potentials" (Deleuze 2014, 322).

It means, it is via the living organism that we are capable of conceiving the cosmos as a living chaos, as a whole that is in the process of individuation. In the realm of physical individuation, the sole unity or interiority is that of the universe; only the infinite and the chaos can assist in metaphysically comprehending the process. Inversely, an organism, by which a true interiority emerges, provides a stage on which one can witness the process. This view emphasizes the pivotal role of the milieu, both inner and outer, in biological individuation. According to this perspective, an organism can be thought of as a universe in itself, with an inherent independence from its milieu. But as Barthélémy observes, "attributing "true interiority" to the living individual is not the same as making it substantial—fighting against substantialism even being the whole point of Simondon's thinking of individuation" (Barthélémy, 25). This independence is supported by the internal metabolisms that drive self-generation and self-development.[11] But physical individuation is not exhausted in biological individuation, and this makes the so-called biological universe a theater on a large stage, within a much larger universe which entails a similarly complicated organism.

Furthermore, a distinction between physical individuation and biological individuation lies in the fact that, in the former, the pre-individual state and the individual are somehow distinct or separated. However, in the case of an organism, the pre-individual state remains inseparably interconnected with it. Physical individuation results in the creation of an object that is complete in itself. In contrast, biological individuation generates an individual that continually individuates itself, never attaining the stability of a physical individual. In other words, it is insufficient to describe a living being in terms of the concepts of structure and content, which are derived from the physical notions of form and matter.

This renunciation must be applied retrospectively to the physical domain, which is Simondon's objective in the first chapter of "Physical-Biological Individuation," leading us to the allagmatic nature of the entire realm of individuation. On one hand, as Simondon notes, "In physics, heterogeneous transductive characteristics only appear in the margins of

[11] "It could then be supposed that the external richness of the relation to the milieu is equal to the internal richness of organization contained in an individual" (*Individuation*, 170).

this physical reality; on the contrary, interiority and exteriority are everywhere in the living being" (172). On the other hand, if we consider the entire domain of physical reality, including what is referred to as chaos, we find that it also contains interiority and exteriority throughout. In simple terms, while the individuation of a specific physical object can be explained by the modulative principle, leading to the formation of a self-contained individual, the broader physical domain, governed by the physical paradigm, can be described as the theater of individuation. The physical chaos individuates itself like a living organism. Given this premise, the next step involves returning to the individual living organism, with the aim of dismantling its organic unity, of machinizing it, all the while blending the internal and external milieus into a machinic whole.[12]

Psychosocial (trans)Individuation

Biological or vital individuation gives rise to the formation of self-preserving organisms that actively carry their history with them, resulting in the constitution of psyches. According to Simondon, this involves a phase change, meaning that the structure and rhythm of individuation in this context do not exhibit any direct dependence on physical-biological individuation, although they are subjected to the same process. We move from one level of individuation to another, although "the same being can exist on different levels" of individuation (167). It is because of these phase changes that we can discuss various types of reality. Now, in the case of the psychical, "the psychical and the vital are not distinguished as two substances or even as two parallel or superposed functions; the psychical intervenes as a slowing down of the individuation of the living" (177). Thus, it is a matter of the tempo of individuation in relation to the individual and pre-individual reality, the entity and the milieu. What is then individuated is the conscious individual, the

[12] Audronė Žukauskaitė in her chapter on Simondon asserts this fact in a concise way although she does not derive from this a machinic account of organism: "the structure of a complex organism is like a multiple folding where the internal milieu is enfolded into an external milieu, which in its turn becomes the internal milieu to be enfolded into a more complex external milieu" (Žukauskaitė, 29).

subject, whose act of individuation is perception, with its pre-individual state appearing as its sense and emotion. The individuation of the psyche can be understood in terms of the pre-individual sense, emotion, and affect, as well as the individuated consciousness that perceives objects.

Simondon defines psychical individuation in terms of "psychical tension," which indicates the "degree of metastability" in the realm of psychical individuation (260). He describes this concept again using the terms form and information. The degree of metastability indicates the amount of information that each form contains. The affective information is what conditions a psychic form from within. The individual subject acquires its form through its ability to perceive, but this perceptive form is based on an affective preindividual flux that provides an abundance of information.[13] On the other hand, sensation indicates the relation of the psychic form with the external preindividual flux. Psychical individuation represents the manner in which internal and external information gives rise to the formation of the psychical individuals and societies of them. These two informational directions constitute the subconscious ground of individuated consciousness (273). According to Simondon, this structure is shared by all living beings. Hence, the distinction between humans and other animals lies not only in the extent or restriction of their affective informational world, but more crucially in how they create form from that information, i.e., how they transform sensation and affect into perception and emotion as exterior and interior unities or universes. The shift from information to form involves a move from diversity to unity and at the limit the formation of species, types and worlds (289). Interesting for phenomenologists, Simondon's psychical-collective individuation implies that the individuation of the internal unity is parallel to the individuation of the external unity.

Therefore, a trans-individuating operation is at work here that links the psychical individuation to the collective individuation and the

[13] Simondon compares information theory with the theories of form such as Gestalt theory. By information theory he means a branch in the philosophy of science that considers the transmission of signals as a solid part of the system in question. Simondon's examples are relatively classic and he does not refer directly to the theory of quantum information, although his arguments are in a similar direction (*Individuation*, 265–7). A good account of the relation between information theory and quantum mechanics is provided by Herbert S. Green in *Information Theory and Quantum Physics Physical Foundations for Understanding the Conscious Process* (Green 2000).

individual history to the social history. Again, it is the trace of the pre-individual that makes the psychical individual metastable and forces it to connect its individuation to collectivity:

> It could be said that biological individuation does not exhaust tensions, which have aided in its constitution: these tensions pass into the individual; they pass into the individual from the pre-individual, which is both milieu and individual: it is precisely based on this position of the unresolved in man, within this not-yet- individuated charge of reality, that man seeks out his fellow man to form a group in which he will find presence through a second individuation. (339–340)

The "not-yet- individuated charge of reality" in the individual constitutes the source of its collectivity. This is to say that, although we are dealing here with a phase-change, but the new phase of individuation functions within and over the physical-biological individuation. If the psychical-social marks the realm of Man, Simondon insists that, "the vital is the vital that includes Man, not the vital without Man; it is the vital up to Man and including Man; there is the whole vital, which includes Man" (332). Man is already included in the vital in the form of the individual. But no individual is exhausted and the pre-individual charge of reality traverses each psychical individual, rendering it inherently collective. This is the core meaning of transindividuality. The trans-individuating operation describes how physical-biological individuation can separate and transform the psychical metabolism into a collective individuation. The relation of an individual to other individuals and its relation to the pre-individual are in essence the same. The psychic individual and the social individual lose their integrity in the process of individuation:

> The interindividual relation goes from individual to individual; it does not penetrate individuals; transindividual action is what makes it such that individuals exist together as the elements of a system that contains potentials and metastability, expectation and tension, then the discovery of a structure and of a functional organization that integrate and resolve this problematic of incorporated immanence. The transindividual passes into the individual as though from individual to individual. (339)

This is the basis for Simondon's critique of both psychologism and sociologism, which involve the segmentation of interior and exterior groups (330–331). The issue with psychologism lies in its tendency to isolate the psychological individual from its collective essence. On the other hand, sociologism treats social relations as external to psychical individuals. A similar line of reasoning extends to the concept of interiority within social groups compared to other groups or societies. Transindividuation deconstructs this dichotomy of interiority and exteriority. Consequently, Simondon critiques the conventional understanding that views the interiority of a group as external to the interiority of the individual psyche. This is why, for him, it is undecided whether to take the ordinary known individual organism as the interiority or the colony in which the individual takes part as an organ.

One of the main aims of Simondon's account of the collective is to prevent any reduction to mere intersubjectivity. The prefix "trans-" takes the place of "inter-". The latter is used to refer to the relationship between fully constituted individuals, while the former marks what traverses the individual while individuating it. In other words, the term transindividual does not imply the relation between psychic individuals in a society but that which collectivize the psychic individuals through the pre-individual register in each of them. As Jason Read declares, "Transindividuality is not intersubjectivity: it is not a relation between constituted subjects, but rather a relation between the constitutive conditions of subjects" (Read 113). This transindividual collectivity must be considered the source and core of sociality. The formation of a group, far from being a joining together of individualized persons, is described by Simondon with the term "syncrystallization" (*Individuation*, 333). In addition to highlighting the paradigmatic role of physical individuation in psychical-collective individuation, this term signifies that the formation of the psyche and that of a group are intrinsically linked processes. What is observed in (post) structuralist psychoanalysis, where subjectivity is recognized as intersubjectivity, finds a deep resonance in Simondon's perspective: the individual is the transindividual and collectivity forms the core of the individuation process from its inception: "The individuation that gives birth to the group is also an individuation of grouped individuals; without emotion, without potential, and without preliminary tension, there can be no group individuation; a society of monads cannot exist" (332). This also means that structuration and operation are the same thing. The only thing that makes individuation a structuration is its inherent collectivity. In this sense, "Individuals are group individuals" (334). Moreover, as noted by Simondon, the significance of conceptual communication in inter-subjectivity yields to the prominence of "emotion" in transindividuation. Emotion, which can be viewed as the synthesis of affection and sensation, serves as the bridge connecting the individual to both their inner preindividual core and external individuals simultaneously.

Machinic Ontogenesis

The ontological approach in Simondon's philosophy of individuation, within the domain of psychical-collective reality, implies an anti-vitalist view towards the realm of human life, serving our machinic approach. The trans-individual operation that makes the psychical inherently incorporated in collectivity issues from a pre-individual force that, according to Simondon, is not vital:

> This force is not vital; it is pre-vital; life is a specification, a first solution, complete in itself by leaving a residue outside its system. It is not as a living being that man carries with him enough to individuate spiritually, but as a being that contains within itself something pre-individual and pre-vital. This reality can be called transindividual. Its origin is neither social nor individual; this reality is deposited in the individual, carried by the individual, but it does not belong to the individual and is not a part of the individual's system of being. (340)

Simondon describes human alleged spirituality in terms of something that, far from being related to individual life, issues from a pre-individual and pre-vital transindividuality or collectivity. The pre-individual at the center of psychic individuality underlies collective individuation, but this entails a primordial collectivity that is prior to vital substances. According to Simondon, there is a collective pre-individual that is logically and ontologically prior to the living subject as the principal psychical element. Subjectivity is inter-subjectivity, but inter-subjectivity is rooted in the pre-individual transindividuation (syncrystallization) at the heart of the subject. As Žukauskaitė asserts, this also can be called the trace of "nature" at the heart of human individual: "the individual should be defined not as a substantial being, but as a being which carries within itself something non-human (nature, the pre-individual charge), and which can resolve its tensions only by going beyond itself towards the transindividual collective" (Žukauskaitė 2023, 35). We have long recognized that humanity cannot be simply defined by its individual forms. However, what Simondon uncovers is that the collective nature of humanity underpins the fundamental role of non-human elements in our existence. And Simondon emphatically asserts that the non-human component is inherently non-vital as well.

As mentioned, the collective is the transindividual itself, but it can generate individuals in different levels and modes. Therefore, one can distinguish, for example, between (human) social collectivity, animal collectivity, bacterial collectivity, or quantum collectivity. Nonetheless, the fact is that they all are modes of one and the same being that individuates itself. Collectivity belongs to individuation itself in the entire domain of being per se. In this sense, the collective is much more profound than the inter-subjectivity of conscious agents. It has an ontological root. In this way, Simondon bases subjectivity, in all of its complexity, on a pre-vital metabolism. The result is transindividual collectivity in any domain, which does not entail the relation between individuals, but rather the transindividual relations that put the individual in relation with its individuation. The relation between individuals constitutes only the conscious surface of the collective. What constitutes the subject is not the subjective core (self-consciousness), but rather that which dissipates subjectivity through its transindividuation. As Combes observes, "… the presence of the collective within the subject in the form of an unstructured preindividual potential, constitutes a condition for the relation of the subject to itself" (Combes 39). If a subject is defined as a stable conscious psychic being, the transindividual relation between subjects would be a genetic relation between their constitutive parts that are always under construction, and the relation is the same as their becomings. Combes maintains, "Thus these shares of nature, of real potential, are what link individuals to one another in the collective; it is because of them that constituted individuals can enter into relation with one another and constitute a collective" (Combes 47). Only what is not individuated in individuals is responsible for the relationship between them, if this relationship is supposed to be substantial, and not accidental. The substantiality of a relation between individuals stems from the fact that they are separated from a universal chaos and become individuated through transindividuation.

The problem of our social life finds its solution in the source of our existence, which lies in how we are constituted from undifferentiated matter, or what is referred to as chaos. We are produced collectively in nature, in the same manner as all entities, from bacteria to mammals, are generated collectively. Society has a (meta) physical ground. At the level of this profound collectivity, the differentiation between species that Simondon discusses is irrelevant because even the species that are not forced by their milieu to lead a collective life are not exceptions to it (337). They are collective as well.

It is at the level of this primordial collectivity that philosophy should seek the source of human spirituality and historicity, or the realm of so-called culture. The Simondonian chaotic (or contingent) physicalism can lead us to conclude that nature is not simpler than culture. Precisely at this point, Simondon's philosophy of nature (his philosophy of individuation) and his philosophy of culture (his philosophy of technics) join together. He brings these two realms into the same unified field. Both realms constitute different aspects of a single ontogenesis. Therefore, the concept of pre-individual collectivity in *Individuation in Light of Notions of Form and Information* addresses the central problematics posed in *The Mode of Existence of Technical Objects*. In this regard, we disagree with Combes who sees Simondon's naturalist attitude and his technical thought as "*two incompatible tendencies of thought*" (Combes 57). Simondon's philosophy of technics should be regarded as an extension of his concept of psychical-collective individuation, encompassing the cultural-historical dimension of human life. This is obvious from the first lines of *The Mode of Existence of Technical Objects* where Simondon claims that "The opposition drawn between culture and technics, between man and machine, is false and has no foundation; it is merely a sign of ignorance or resentment" (*Mode*, 15). Technical objects are not merely objects made by humans to be utilized. They are the theater of natural autonomy, as Simondon describes "a world of technical objects as mediators between man and nature" (15). Technical objects, which are synonymous with machines, possess a form of autonomy and operate in an allagmatic manner. They inherently function as systems comprising various interconnected functions. It is in the mode of existence of technical objects that we can search for the pre-vital allagmatism that constitutes human cultural life because the way a technical object is distinguished from a non-technical object is similar to the way an organism is different from a mineral. This is why Simondon goes as far as to equate the machine with life itself (21). In *The Mode of Existence of Technical Objects*, he strives to demonstrate that within any machine there is a human locked up; in *Individuation*, he indicates that within any human being there is a (natural) machine locked up. By transductively extending the allagmatic principle to the entire realm of existence, we can ascribe a machinic ontology to Simondon's thought, where the only agency is the machinic ontological agency that links the natural and the technical. This agency was mistakenly delegated to humans for a short time (the era of subjects). The so-called industrial revolution marks nothing but a return to the natural agency (81). The mode of existence of technical objects must be explored through the history of technicity, which is intrinsically a part of natural history.

Simondon expounds upon this concept through his notion of the technical milieu, which he also refers to as the "associated milieu" (59). The technical object operates within the natural sphere, engaging with its physical, chemical, geographical, and biological surroundings. What's crucial here is the technical object's capacity to make discerning choices, selecting specific elements from its environment to construct its unique milieu. This underscores a machinic agency operating precisely at the interface between the technical object and nature. This agency renders the technical object a quasi-independent entity that carries its associated milieu with itself, constituting a self-conditioning system that can transform the pre-individual contingency into a kind of machinic creativity.

Humans exist "at the same level as machines" (125). The notion of interaction with the associated milieu holds central significance not only for technical objects but also for all biological entities, including humans. In this context, creativity does not stem from subjectivity but rather from the emotive latency of human life which is inherently collective and transindividual. Man is machine but of course not in a utilitarian or mechanistic sense. Simondon's efforts to disavow machines from a utilitarian and mechanistic conception is compatible with his efforts to link or even identify men and machines. Human creativity finds its foundation in the machinic creativity that governs the entirety of ontogenesis. This principle also applies to human society, as depicted by Simondon as an autonomous, non-organic agency. Simondon's naturalistic (or ontological, or anti-anthropological) approach sheds a new light on this thought. Society exists at the same level as machines because it has its own autonomy and agency, creativity and productivity, and phase of individuation independently of the individuals involved. It gains its autonomy from the pre-individual force that enables individuation to operate.

To fully understand technical objects, it's necessary to consider them in their historical context and evolutionary metabolism. There is no way to grasp the essence of technical objects except by studying their ontogenesis. Technical history represents the virtuality of machines, even though in their individual existence, they may appear to be entirely actual. As Thomas Lemke declares, "In Simondon's work technology is not a material object or the product of thought, but rather an incessant process of adjustment and repair" (Lemke, 94). While technical objects can be seen as self-conditioning systems akin to living organisms, they are not self-contained individual entities crafted solely in alignment with the intentions of their creators. Instead, they possess their own process of individuation that unfolds within the context of both natural and cultural history. This allows them to possess a fundamental sociality that permeates the entire realm of existence.

In this context, Simondon's concept of machinic collectivity, viewed through the lens of technicity, is framed by the notion of the network. A key outcome of Simondon's philosophy of the trans-individual is that reality is inherently a networked reality. In this sense, a network is technical but not manipulable. Thus, technical objects are, by nature, networks, distinguishing them from non-technical objects. First of all, this is because the unity of technical objects is a transductive unity: "Technical objects are multiple and fragmentary by nature; technical thought, enclosed within this plurality, can progress, but only by multiplying technical objects, without being able to recapture the primitive unity" (186). It is a networked unity, and this is precisely what makes technical objects unmanipulable: "one cannot change the network, one doesn't construct a network of one's own; one can only connect to a network, adapt to it, participate in it; the network dominates and frames *[enserre]* the action of the individual, it even dominates each technical ensemble" (229). The unmanipulability of technical objects, which arises from their networked nature, represents their autonomy.

Here, we want to emphasize that through the mediation of technics, the distinction between human reality and natural reality becomes blurred. The human collectivity is rooted in what Simondon calls "the magical network of the universe":

> This act becomes an outstanding point of the network of human life integrated within the world; from this outstanding point a higher kinship with others is created, reconstituting an analog of the magical network of the universe. (192)

Transductive unity, as the philosophical core of Simondon's thought, which is a networked unity in the terminology of *Mode*, permeates the individual and breaks it down into organs, so to speak. Each individual plays the role of an organ for a collective individual while also being a network of other individuals: "The relation between singular being and the group is the same as between the individual and subindividuals. In this sense, we could say that there is a homogeneity of relation between the different hierarchical scales of the same individual and similarly between the group and the individual" (*Individuation* 170–171). Human technicity is an extension of human collectivity and historicity, or in other words, human culture. Similarly, there are other forms of technicity that can be considered an extension of a specific animal culture. And in turn, any collectivity finds its condition of possibility in the inherent ontological collectivity. The networked nature of technics mirrors the networked nature of life and existence, and thus serves as a reflection of it.

In section II of part II in *Individuation*, called "Specific Form and Living Substance", Simondon describes how coelenterates demonstrate that "Life can exist without individuals being anatomically and physiologically or merely physiologically separate from one another" (180). He explains how these living beings form a colony in which each individual functions as an organ:

> In certain colony formations, the individuals manifest a differentiation that winds up in some sense transforming them into organs: some have a nutritive role, others have a defensive role, while others have a sexual role, and it could be claimed in some sense that veritable individuality is transferred to the colony if an impregnable residue of individuality didn't remain in the differentiated beings that compose the colony. (181)

Simondon argues that a colony is not an organism; therefore, the death of a certain number of individuals in a colony does not threaten its existence, unlike in an organism where the death of some organs can be fatal. However, one could argue that if the defensive part of the colony were to die, it could be deadly for the entire colony. Simondon acknowledges that a living individual is mortal, while a colony is not an individual in the same sense. In terms of the role of an organ, a colony is a body without organs and is never truly individualized:

Because of this, it should be said that amoebae, as well as a large number of infusoria, are not strictly speaking veritable individuals; these beings are capable of regeneration by exchanging one nucleus with another being, and after a period of time they can reproduce by dividing into two parts; certain holothuria can also divide into a plurality of segments when conditions of life become poor, each segment afterwards reconstituting a complete unit, i.e. a holothuria similar to the previous entity. In this case and properly speaking, there is no distinction between the individuals and the species; individuals do not die but divide (181).

This exemplifies a pre-individual system of existence. A few pages later, Simondon insistently emphasizes that it is not important to define or characterize what an individual is: "In a word, what is an individual? We respond to this question that we cannot rigorously speak of the individual but of individuation; we must come back to the activity, to the genesis, instead of attempting to grasp the fully formed being in order to discover the criteria by means of which we will know whether or not it is an individual. The individual is not a being but an act, and the being is an individual as an agent of this act of individuation through which it appears and exists" (208–209). The emergence of life is a pivotal phase in the process of individuation because it gives rise to the formation of organic individuals that can be considered true individuals. The example of the coelenterate colony challenges the distinction between the organism and the simple unorganized object. The so-called collective unity is not organic, if human collectivity gains its substantial core from the magical network of the universe. The unity of societies is not organic, it is machinic. This is equivalent to saying that the transindividual lines within the collective must be considered independently of the formation of organic individuals produced by the individuation of life. Hence, the "transindividual universe" is the only existing universe. It is the same being which suffers and enjoys phase-change from the physical-organic individuation to psychic-collective individuation.

This leads us to the consideration of networked reality which is as much machinic as it is natural, and as much human as it is machinic. The source of human sociality is the same as that of colonies and networked technologies, although it results in a different type of sociality. It is machinic-natural. Jason Read declares this point in his study: "These two problems, the relation of the individual to technology and of the individual to society, constantly intersect while obscuring each other" (Read 106); or, "To adequately grasp technology, it is necessary to adequately grasp collectivity, and vice versa" (Read 107).

This can have several consequences for the notions of human life and science. First of all, it is only through a machinic ontology that human sciences can attain the status of a science. No human science can be considered scientific if culture is not studied in light of natural phenomena, and human technics are not studied in light of natural technologies. However, it is important to emphasize that this does not entail objectivism, physicalist or biological reductionism, as it takes physics and biology in terms of absolute contingency. It is only through a probabilistic physics and a non-vitalist biology that a scientific ground for humanities can be established. On the other hand, it does not apply human subjective principles to physics and biology, as it relies exclusively on natural sciences in order to make sense of human culture and technical life (it is not a subjectivism). The scientific contingency is not derived from a conscious human notion of freedom. This scientific approach aims to provide a machinic account of human freedom, and only under these conditions can humanities focus on collectivity instead of individualistic consciousness. Read finds a similar perspective in a comparison between Simondon and Marx:

> Simondon's general schema is similar to Marx's general understanding of the development from tools, governed by the schema of an instrument working on an object, to machines, which reduce man and 'his tools' to a part of the machine, a conscious organ of a greater machinery. (Read 104)

In the collective transindividuality, human life, nature, and technology converge. Technological advancements give rise to new forms of collectivity. How can transindividuality, as the interplay between individuality and collectivity, reshape our understanding of technology? The answer lies in distinguishing between form and information, and in de-formalizing the concept of technology. In this regard, and in the context of Simondon's idea of transindividuality, Read concludes that there is no individual machine: "This undermining of the individual person is further called into question by the individual status of the machine. It is notoriously difficult to define the individual device, or machine; beyond the so-called simple machines, any technological device is made up of multiple components. It is often difficult to draw a line of demarcation between one device and others" (107–108). As it is clear, Jason Read reads *The Mode of Existence of Technical Objects* as the next step in Simondon's philosophy of individuation after the psychic-collective individuation; collective individuation gives rise to the technical-machinic individuation: "This shift and enlarging of the initial problematic is not a radical break; the general problems of technology and alienation persist throughout Simondon's writing on individuation" (108). Hence, the machinic transductive unity. Machinic ontology represents the ultimate goal of Simondon's ontological approach; his philosophy culminates in the thought of machinic ontogenesis. This continuous approach implies the extension of the "re-naturalization of humans" to technologies and machines: If reading human collective and cultural life in light of physical, organic, and psychic individuation results in putting this collectivity in the heart of nature and making an inherent link between nature and culture, then machinic ontogenesis would also imply a naturalistic view towards machines.

References

Bachelard, Gaston. *The Formation of the Scientific Mind: A Contribution to a Psychoanalysis of Objective Knowledge*. Trans. Mary McAllester Jones. Manchester: Clinamen Press, 2002.

Bardin, Andrea. *Epistemology and Political Philosophy in Gilbert Simondon: Individuation, Technics, Social Systems*. Springer Science+Business, 2015.

Barthélémy, Jean-Hugues. *Life and Technology: An Inquiry into and Beyond Simondon*. Translated by Barnaby Norman. Meson Press, 2015.

Combes, Muriel. *Gilbert Simondon and the Philosophy of the Transindividual*. Translated by Thomas LaMarre. The MIT Press, 2013.

Deleuze, Gilles. *Difference and Repetition*. Trans. Paul Patton. Bloomsbury Academic, 2014.

Green, Herbert S. *Information Theory and Quantum Physics: Physical Foundations for Understanding the Conscious Process*. Springer, 2000.

Nancy, J. L. The Deleuzian fold of thought. Trans. T. Gibson and A. Uhlmann, in P. Patton (ed.) *Deleuze: A Critical Reader*, Oxford: Blackwell, 1996.

Read, Jason. *The Politics of Transindividuality*, Leiden and Boston: Brill, 2016.

Simondon, Gilbert. *On the Mode of Existence of Technical Objects*. Translated by Cecile Malaspina and John Rogove. Univocal Publishing, 2017.

Simondon, Gilbert. *Individuation in Light of Notions of Form and Information*. Translated by Taylor Adkins. The University of Minnesota Press, 2020.

Toscano, Alberto. *The Theatre of Production: Philosophy and Individuation between Kant and Deleuze*. Palgrave Macmillan, 2006.

Žukauskaitė, Audronė. *Organism-Oriented Ontology*. Edinburgh: Edinburgh University Press, 2023.

4

Catherine Malabou and the Plastic Machine

Catherine Malabou's philosophical work draws a parallel between the brain's plasticity and the structure of social and political life. She investigates the ethical and political implications of the epigenetic fact that the criteria for formation only emerge after the process of formation. In other words, now that we know our brain has no fixed structure governing our decisions, how should we navigate our social lives? In a transitional phase of her work, she also explores a parallel between the human brain and natural history, drawing on the work of historians like Daniel Smail and Dipesh Chakrabarty. Unfortunately, she does not elaborate on this issue, and apart from a few lectures given between 2015 and 2017, there are scarcely any references to this topic in her published work. This chapter aims to examine this transitional phase of Malabou's work, exploring how the issues addressed during this period might shift her socio-political thought toward an ontological framework. In other words, our focus is on how the epigenetic nature of the human brain can introduce the concept of natural history as a plastic brain. During that period, Malabou was interested in the relationship between the brain and natural history. Our focus, however, is on the ontological or metaphysical idea of conceiving natural history as a form of brain. The relationship between

M. Parsa, *Machinic Ontology*, https://doi.org/10.1007/978-3-031-99561-3_4

natural history—what happens to us—and the history we create mirrors the connection between a brain that happens to us and a brain that we shape. If the brain is an ongoing event, then natural history can be viewed as the ontological extension of the brain.

This marks an ontological tendency in Malabou's work that can be seen as a continuation of a trajectory in twentieth-century philosophy of science, as represented in the works of Georges Canguilhem and Gilbert Simondon: with Canguilhem, human problems are considered part of the same processes that facilitate interactions between an organism and its milieus. Canguilhem takes a significant step in situating human problems within biological nature when he writes in the Introduction to *Knowledge of Life*, "Doubtless, the animal cannot resolve all the problems we present to it, but this is because these problems are ours and not its own. Could man make a nest better than a bird, a web better than a spider?" (Canguilhem 2008, xviii). With Simondon, this biologism extends to an ontological level by introducing physics as the paradigm for the entire philosophy of individuation (See Combes 2013, 12–14). Building upon the work of Canguilhem and Simondon, Malabou discovers the brain not just as a neuroscientific information-processing machine, but also as a philosophical tool for developing a new materialism that encompasses all of nature, although unlike Simondon, who seeks the ontological paradigm in physics—particularly the concept of metastability—Malabou looks for the ontological paradigm in natural history, ecology and geology. In a diversity of ways, Malabou attempts to give a philosophical concept of brain which transcends the human brain or the subject matter of neuroscience. If German idealism centers upon the identity of *Geist* (spirit) and nature, Malabou's philosophy can be considered a development in that philosophical tradition that centers upon the relationship between brain (a neurotic system) and natural history (a cosmic system). She does not go so far as making an identity between these two realms, although her work directs us towards such an identity. If Hegel's contribution in philosophy can be translated as making the Kantian transcendental a historical epigenesis, Malabou's work is no less than a materialization of this epigenesis in two directions of neuroscience and natural history. Within this context, the present chapter aims to focus on the notion of "the brain of history" which is introduced, but not sufficiently elaborated, by Catherine Malabou. Our aim is to introduce this notion of the brain of history as an example of plastic ontologies, as appears in Joshua Schuster's comment: "Malabou asserts our epoch is better described as one of plastic ontologies and partially modifiable histories and forms" (Schuster 2018, 186).

Therefore, the goal is to extend the relationship between neuroscience and philosophy in Malabou's early work to the relationship between neuroscience and natural history. Before jumping into Malabou's work on deep history let us examine the problematic from two points of view, namely a static cosmological view and, an evolutionary biological view. For the first case, let us deal with a study regarding the relationship, or parallelism, between neuroscience and cosmology. In this regard, Malabou's work can be considered an attempt to philosophize the results of a study that Franco Vazza and Alberto Feletti ran. The study is titled "The Quantitative Comparison Between the Neuronal Network and the Cosmic Web". This study aims to provide a quantitative basis for drawing parallels between the intricate connectivity observed in neuronal networks and the cosmic web structure of galaxies.[1] Vazza and Feletti, childhood friends who respectively specialize in cosmology and neuroscience, embarked on an endeavor to compare the network of neuronal cells in the human brain with the cosmic web of galaxies. Through their collaborative effort, they uncovered intriguing quantitative similarities between these disparate systems. While autopoietic theory, as demonstrated by figures like Maturana and Varela, has already established that living systems are cognitive systems, influencing Žukauskaitė to formulate an organism-oriented ontology (Žukauskaitė, 8–9), the study conducted by Vazza and Feletti propels this discourse to a more advanced stage, taking them as analytically stable. Their research contends that even non-living systems, such as the cosmic web, could be deemed cognitive systems.

Building upon recent neurological discoveries, Vazza and Feletti interpret the human brain as a network of connectomes, intricate configurations of neural connections that give rise to complex cognitive functions.

[1] https://www.frontiersin.org/articles/10.3389/fphy.2020.525731/full.

Following their analysis of the neuronal network, they draw a comparison with the universe as described by the ΛCDM model (Lambda Cold Dark Matter). This model, rooted in the theory of General Relativity, incorporates gravity's effects on both ordinary and dark matter, as well as the presence of "dark energy", which represents the anti-gravitational energy associated with empty space. By juxtaposing the intricate interconnections within the neuronal network with the cosmic web structure governed by the ΛCDM model, they reveal intriguing parallels between the complexity of neural networks and the vast cosmic fabric. The concept of the cosmic web, as explored by Vazza and Feletti, pertains to the distribution of matter density fluctuations within the universe. This distribution manifests in various structures, including larger groups or clusters of galaxies, filaments, matter sheets, and voids. These components form an intricate networked structure that spans the cosmos, revealing the underlying organization and interconnections within the fabric of the universe. On this basis, Vazza and Feletti seek to establish a comparison between the cosmic web and the neuronal network, which they define as a "multiscale structure". Within the neuronal network, neurons cluster together to form circuits, columns, and various interconnected functional areas. To conduct their research, Vazza and Feletti selected samples from slices of the neuronal cortex. By focusing on this specific region of the brain, they were able to analyze and examine the intricate patterns of neuronal connectivity within the cortex. These selected samples served as representative snapshots of the complex network present in this crucial brain region, allowing them to draw comparisons with the cosmic web structure.

What insights can we gain from this comparative study? What we seek here is not merely parallelism or analogy, but rather the guiding principle for discovering the elements of an ontology. Based on this, the principle can be articulated, alongside other similar concepts, through connectionism: The functioning of a network is independent of the material substance of its nodes. The information generated by a network is purely networked information. This perspective leads us to an ontological concept of thought, reflecting Gilles Deleuze's peculiar structuralist approach in *Logic of Sense*. There, thought is understood as the incorporeal effect (sense) of corporeal actions and passions, a metaphysical surface built upon the physical surface (Deleuze 2015, 228). In the concluding lines of his chapter on Thought, Deleuze introduces the brain not merely as a corporeal organ but as the totality of the metaphysical surface on which all events are inscribed and symbolized—a thinking universe. He writes, "The brain is not only a corporeal organ but also the instructor of another invisible, incorporeal, and metaphysical surface on which all events are inscribed and symbolized" (ibid, 230).

While comparing the brain to the cosmos is a spatial and static approach, Malabou suggests that a more fruitful direction is to view the brain as identical with natural history.[2] How can natural history depict a neuronal network? It would be a materialization of Hegel through the Deleuzian idea that "things make sense through their genesis" (Parsa 2023, 2). It could be translated as: a material system thinks through its evolution. This leads us to the second, evolutionary perspective that Malabou explored in her analysis of Charles Darwin as the seminal figure who introduced contingency into natural history. Malabou links the contingent generation of species in biology (natural selection) with the contingent generation of individuals in history (social selection) as two interrelated examples of plastic formation and information (Malabou 2015, 49). This notion of historical contingency leads Malabou to the concept of emergence from nothing, *ex nihilo*. This is precisely what makes the idea of deep history significant: deep history differs from prehistory in that it denies any causal agency. In this regard, the foundational idea of deep history suggests that what is often regarded as historical determinism (the orthodox reading of Hegel) is actually built upon historical contingency. As Malabou asserts, "Natural selection transforms the contingency of the former into a necessity" (50). Natural selection regulates the formation of forms, but the crucial point is that, by incorporating the idea of contingency, Malabou offers a

[2] Alberto Toscano explains, and critiques, this "exchangeability" as the common property of ontology and capitalism. He refers to Heidegger's notion of ontology, and Malabou's text on Heidegger, *Le change Heidegger*. Toscano contends that it is the cooperation of metaphysics and capitalism that leads to the logic of "everything is equal to everything, any being can be exchanged for an other" (Toscano 2015, 94). This means that for Heidegger, being is indifferently exchangeable. In short, according to Toscano, we are not in the realm of ontology, but rather metaphysics and capitalism.

non-mechanistic interpretation of Darwin. This means that one cannot predict the outcome of natural selection simply by comparing struggling species and determining which is stronger—a "confused Darwinism and Malthusianism" (51).

What can natural selection teach us about the functioning of the brain? Malabou insists that natural selection is "without intention" and that "In nature, selection is unconscious" (52). And she maintains, "As soon as selection becomes an intention of selection, which presupposes pre-defined criteria, certainly programmed this time, as soon as there is no more naturality or spontaneity in the promotion, the plastic condition is menaced or even inexistent" (52). A brain that decides without intention reflects a nature that selects without a program (and a society that elects without sovereignty, leading Malabou to discuss Althusser alongside Darwin). In short, the only existing plasticity is the plasticity of nature and any plastic brain would be nature as brain.

Plasticity allows us to apply the "topographic network" of the brain to various other networks, including not only cosmic networks but also computer and AI systems (Malabou 2019, 20). This direction was explored by Malabou in *What Should We Do with Our Brain?* (2004) and more recently in *Morphing Intelligence* (2019). As mentioned, Malabou examines the relationship between neuronal systems and other natural or artificial systems. In her later work, Malabou continues the lines of thought presented in the second chapter of *What Should We Do with Our Brain?*, focusing on a comparative study of the brain's plasticity and the structure of capitalist political and economic power, which demands the brain's flexibility for social and political integration. This line of thought can be formulated under the title of "intelligent resistance". We also examine Malabou's notions of the plasticity of artificial and social brains, but only to approach her conception of nature as brain. Therefore, in this chapter, we aim to highlight an aspect of Malabou's work that she, for compelling political reasons, has shifted focus from: namely, "plastic ontologies and partially modifiable histories" (Schuster, 186). Our aim is to explore the notion of the brain, whether it appears as an AI system, a galaxy, natural history on Earth, or a society, and to understand how it thinks contingently.

Anthropocene and History

In a 1978 interview with Bryan Magee, Noam Chomsky asserts that "we are rigidly pre-programmed" and this is the biological basis of our existence.[3] He proceeds to place the enigma of human creativity beyond the reach of scientific inquiry, attributing it to a realm of mystery. He contends that human science lacks the ability to provide insights into matters concerning will, decision-making, reason, or the selection of actions. According to him, these questions continue to reside in obscurity, reminiscent of their status during classical antiquity. In this context, as well as throughout much of his intellectual work, Chomsky maintains a steadfast adherence to Kantianism. In a thoroughly Kantian fashion, he upholds the notion of the rigid foundation of humanity. The history of philosophy following Hegel, along with the advancements of modern science, emerged in opposition to such a belief. The post-Hegelian philosophy and science convey an anti-Kantian message, asserting that there is no fixed foundation for our existence. In this context, Hegelianism represents the integration of the mystery and enigma of creativity into science.[4] The point is that a science incorporating the ineffable would be a plastic science.

Hegel takes the initial step in investigating the realms of human creativity, will, decision-making, and more by attributing them to *Geist* and history. Within a few decades, German philosophy, exemplified by figures like Marx, came to acknowledge that this identification with *Geist* amounts to a critical spiritualism. In this manner, German philosophy introduced the synthesis of historicity and materialism into the annals of philosophical discourse, thereby shaping its trajectory and determining its fate. Historical materialism not only constitutes a fundamental element within the diverse manifestations of Marxism but also serves as the very bedrock upon which Western philosophy and science as a whole are built.

---

[3] https://www.youtube.com/watch?v=PxvSQnmcYLo&t=406s.

[4] This is evident in *Faith and Knowledge*, where Hegel criticizes Kant's and Jacobi's concept of faith as something ineffable, which cannot be integrated into knowledge (see Hyppolite 1997, 10).

This trajectory, among others, prompts Malabou, emerging from the tradition of French (and hence materialist) Hegelianism, to transition from the notion of *Geist* to the concept of the brain within the historical context. Moreover, she emphasizes the biological concept of epigenetics, diverging from Chomsky's scientific foundation rooted in genetics and its Kantian implications regarding the essence of humanity. Her concept of plasticity suggests that the biological, neurological, and transcendental foundations of humanity are inherently changeable, evolving through interaction with the milieu.

In the lectures "Anthropocene: A New History?" (2015) and "The Brain of History or the Mentality of the Anthropocene" (Malabou 2017), Malabou explores the interplay between history and the human brain, offering a reinterpretation of these concepts. While the redefinition is clearly articulated in the first case, it is somewhat implicit in the second.[5] She accomplishes this by establishing a connection between two accounts of deep history, drawing upon two professors of history, Dipesh Chakrabarty and Daniel Smail. The lectures revolve around the concept of the Anthropocene, which characterizes an era where humans assume the role of a geological force. In his article "The Climate of History: Four Theses", Chakrabarty presents the scientific findings on climate change as compelling evidence of humanity's emergence as a geological force, alongside other natural forces. Intriguingly, Malabou draws a comparison between this concept and an incidental reference made by Chakrabarty to Daniel Lord Smail's book *On Deep History and the Brain* (2008). In Smail's work, he offers a (neuro)biological interpretation of deep history, distinct from a geological perspective, by extrapolating the central functioning element of the human brain, addiction, into the realm of deep history and proposing it as its guiding logic. Chakrabarty expresses his disagreement with Smail, criticizing him for reducing the human geological force to a (neuro)biological framework.[6] However, Malabou takes a different stance by asserting that "there is no hostility between these two accounts of history", and attempts "to find a middle point between

[5] https://www.youtube.com/watch?v=eDdTqr-5APg&t=1099s; https://www.youtube.com/watch?v=wJPLGEdRGGc&t=1776s.

[6] "But it is the history of human biology and not any recent theses about the newly acquired geological agency of humans that concerns Smail" (Chakrabarty 2009, 206).

them". It is this intriguing aspect that we would like to delve into and expand upon. But rather than delving into Smail's theory of addiction, we would like to explore the parallels between Chakrabarty's reimagining of deep history in the context of geological agency and Malabou's own contributions to the study of the brain in *What Should We Do with Our Brain?* and *Morphing Intelligence*. But first let us shortly review the problem as it appears in Malabou's lectures on Chakrabarty.

Malabou, implicitly, suggests that the geological history described by Chakrabarty and the biological history portrayed by Smail through the lens of addiction are rooted in the same nature. In the 2015 version of the lecture, Malabou emphasizes the connection she aims to establish between the notions of history and nature. She acknowledges that achieving this linkage requires a transformation of both concepts. Hence, in our interpretation, the concept of the brain should not be confined solely to the human or animal brain. Instead, it should be expanded as a philosophical concept that encapsulates the functioning of geological deep history. This includes both the distant past (where no traces of humanity exist) and the distant future (when humanity has ceased to exist). This expansive notion of the brain, described by Smail in terms of addiction, encompasses the (post-Hegelian, materialist) logic by which geological deep history operates. While Malabou does not explicitly state this, it becomes evident at a crucial juncture in her lectures where she acknowledges Smail's extension of the concept of the brain far beyond mere reflexivity and consciousness. This recognition implies an implicit agreement that the notion of the brain encompasses a broader scope, including the functioning of geological deep history as described by Smail in terms of addiction. In this context, Malabou conceptualizes the notion of the brain as "the intermediary locus between consciousness and the suspension of consciousness". She draws a weak (which is, anthropocentric, instead of ontological) conclusion from her own discussion by claiming that "we have to get addicted to the idea that we are a geological force". In our perspective, a strong conclusion of this discussion would be to explore the logic of deep history comprehensively. This encompasses not only the deep past and deep future, but the entirety of history, including the period of human existence. It transcends human consciousness or limited biological agency and includes humans as a geological force. This can be achieved by adopting Chakrabarty's geological approach and integrating Smail's understanding of history functioning as a brain. Ultimately, Malabou's primary objective was to provide a fresh perspective on history by bridging the longstanding gap that exists in our understanding of the relationship between history and nature. As Malabou asserts, "nature has a history, it is history". Chakrabarty's first thesis in his essay is straightforward and unambiguous. He boldly announces "the collapse of the age-old humanist distinction between natural history and human history" (Chakrabarty 2009, 201). In this context, it is noteworthy to explore the potential of a non-human understanding of the brain, which could effectively supplant Hegel's notion of *Geist*. This paradigm shift requires us to move away from exclusively emphasizing human history and human-centered accounts of natural history based on a rationalist logic, implying an irrationalist reading of Hegel.[7] Instead, it urges us to recognize and embrace the significance of natural history, rooted in principles derived from geological and biological sciences.

Chakrabarty, in *Climate of History: Four Theses*, writes,

> Humans are biological agents, both collectively and as individuals. They have always been so. There was no point in human history when humans were not biological agents. But we can become geological agents only historically and collectively, that is, when we have reached numbers and invented technologies that are on a scale large enough to have an impact on the planet itself. To call ourselves geological agents is to attribute to us a force on the same scale as that released at other times when there has been a mass extinction of species. (Chakrabarty 2009, 206–7)

When viewed collectively and within a historical context, humans transcend being mere biological agents and instead emerge as geological forces. This point not only deconstructs human agency but also signifies a profound transformation in the agency of biological forms when examined collectively and historically. It highlights a shift in our understanding of the impact and influence of biological entities when viewed in a

[7] "Here perhaps we get to the decisive point of Hegelianism, to this torsion of thought through which we are able to think conceptually the unthinkable, to what makes Hegel simultaneously the greatest irrationalist and the greatest rationalist who has existed" (Hyppolite, 102).

broader context of collectivity and historicity. This is a blatant critique of vitalism, reminiscent of the works of Canguilhem and Simondon. Every biological form possesses a geological agency, as it has the capacity to impact the atmosphere, soil, and other components of the Earth's ecosystem. This view contrasts with Chakrabarty's assertion that "Humans have become geological agents very recently in human history. In that sense, we can say that it is only very recently that the distinction between human and natural histories—much of which had been preserved even in environmental histories that saw the two entities in interaction—has begun to collapse" (207). For a historian, the degree of impact that an agent exerts is a significant factor in determining their status as an actor within the narrative. However, from a philosophical standpoint, the degree of influence exerted by the agent does not really matter. Humans have always acted as geological forces and this is not restricted to industrial age and the measurable impacts of global warming. We would like to emphasize Chakrabarty's assertion regarding the fundamental connection between the biological and the geological realms. This thesis draws inspiration from Lovelock's renowned Gaia hypothesis, which posits that the Earth exists as a self-regulating system. Therefore, while we do not dismiss the relevance of natural historical classifications or deny that the Anthropocene signifies a shift in the Holocene, we would argue, in opposition to Chakrabarty and Malabou, that from a philosophical standpoint, it is not particularly meaningful to assert that we are living in the era of Anthropocene (despite its official recognition by authoritative scientists). This is because all geological forces, whether biological or not, continue to operate collectively and concurrently. By asserting that we live in the Anthropocene, there is a certain degree of anthropocentrism present, which contradicts the understanding that humans are just one among various geological forces. It implies a hierarchical positioning of human agency above other geological forces, which does not align with the broader perspective of humans as mere participants within the interconnected web of geological forces. It is understandable that Paul Crutzen and Eugene Stoermer, as scientists, sought to highlight the "growing impacts of human activities on earth and atmosphere" (Crutzen and Stoermer 2000, 17). However, they did not thoroughly investigate whether their proposal is ontologically equitable. This entails considering whether they had access to comprehensive data on other crucial factors of the planet, such as plankton populations, fluctuations in the number of viruses, and so on, to make a meaningful comparison with the effects of widespread fossil fuel consumption by the human population. While not embracing an early-Foucauldian or early-Latourian scientific relativism, one can still, from a philosophical standpoint, differentiate between the actual geological transformations occurring and the official periodization of these changes as determined by authoritative groups of scientists convening in places like Brisbane or Cape Town. We are still entitled to discern between the natural processes unfolding in nature and the categorizations and designations attributed to them within scientific institutions.

The Plasticity of The Brain

Descartes writes, "the brain is in fact just a tissue constituted in a particular way" (Descartes 2004, 142). This particular composition enables the brain to perform activities that cannot be expected from other tissues. In *The Treatise on Man*, Descartes can be seen as a dedicated materialist and monist, rather than a proponent of mind-body dualism. He strives to explain all mental attributes in terms of the brain's material nature. Although the human body is endowed with a soul, which does not affect its movements, it operates like a machine, similar to animal bodies, which lack a soul.[8] However, Descartes differentiates between natural machines, such as human and animal bodies, and man-made machines, such as "clocks, artificial fountains, mills" (99). In his *Treatise on Man*, he argues that, unlike man-made machines, which are composed of solid parts, natural machines consist of fluxes (104–105). In *Principles of Philosophy*, he further asserts that natural machines are composed of tiny particles that are not directly perceptible to the senses (Descartes 1982, § 201, p. 284).

[8] This reading of Descartes is owed to Georges Canguilhem in *Descartes et la technique* (1937): "The radical distinction between soul and body, thought and extension, implies the substantial unity of matter, whatever its form, and thought, whatever its function" (Canguilhem 1994, 227).

Can the human brain be considered a natural machine in the Cartesian sense? Might this Cartesian conception help in understanding natural history as brain? In the opening page of *What Should We Do with Our Brain?* (2008), Malabou presents us with the intriguing prospect of reconciling history and the brain, leading to the emergence of a meaningful "identity": "It's not just that the brain has a history – which is sometimes confused with that of its constitution as an object of the sciences – but that it *is* a history" (Malabou 2008, 1). Despite this promising start, Malabou does not actually discuss this identity between the brain and history. Instead, she focuses on the brain's plasticity as its historicity. Indeed, what Malabou elucidates in this book is the idea that the brain has a history, rather than being identical with history itself. It is disappointing that Malabou restricts her discussion to the anthropocentric domain of the human brain's work, without venturing into the context of natural history. However, her discussion is intriguing and paves the way for considering natural history as the brain. Stating that our brain is plastic implies that it lacks rigid determination and centrality, which stands in direct contrast to the Kantian perspective we associated with Chomsky, and aligns however with the Cartesian and Hegelian views. Plasticity "means at once the capacity to *receive form* (clay is called "plastic", for example) and the capacity to *give form* (as in the plastic arts or in plastic surgery)" (5). The brain is not a rigid structure but is "truly living" (7). This is the particular way in which the brain tissue is constituted. Moreover, our brain does not serve as the commanding center of our activities. This is a wrong image which is subject to extensive criticism by Malabou. In this regard, the historicity of the brain signifies its dynamic and experimental nature. Malabou stresses that plasticity should not be confused with mere flexibility. It involves a creative resistance to domination and a refusal to be passively shaped. This perspective sometimes leads her to investigate the role of the brain and nervous system in relation to capitalism.

What we find particularly pertinent in Malabou's *What Should We Do with Our Brain?* is her contrasting view of plasticity versus machinality. According to her view, machines lack plasticity and are characterized by rigidity. This represents a conventional interpretation of the concept of machines. However, the universality of the ontological approach suggests that even machines can be seen as possessing plasticity. The concept of the plastic machine represents the intersection where the brain and nature as history (the geological deep history) converge. In the section titled "The End of Machine Brain", Malabou offers criticism of "the well-known technological metaphors that have been used to characterize brain functioning". She critiques these metaphors for their tendency to assign a central and commanding role to the brain (33). The brain is not like a central telephone exchange, as Bergson once suggested, nor can it be compared to a computer. The second metaphor discussed by Malabou appears to be captivating. Those who draw parallels between the brain and computers tend to attribute a programming function to the brain:

> The computer and the brain would in the end both be 'thinking machines', that is, physico-mathematical structures endowed with the property of manipulating symbols. The discovery of the plasticity of brain functioning has rendered such a comparison moot. Plasticity invalidates not the analytical or explicative value of the mechanical paradigm in itself-a paradigm that is, to a certain extent, indispensable for thinking about brain function-but rather the central function habitually associated with the computer and its programs. Opposed to the rigidity, the fixity, the anonymity of the control center is the model of a suppleness that implies a certain margin of improvisation, of creation, of the aleatory. (35)

What if the brain could be understood as a thinking machine that thinks outside the confines of the analytical or mechanistic paradigm? Indeed, Malabou devotes a substantial portion of her work to elucidating how the brain functions by outlining what it is not. However, she provides very limited exploration into how the brain thinks. Through her deconstruction of the classical image of the brain and the decentralization of its functions, Malabou paves the way for a shift towards a machinic depiction of the brain. This new perspective portrays the brain as a nature-machine, characterized by plasticity according to her terminology. Throughout her work, Malabou consistently challenges the human-centric view of the brain, yet she stops short of fully placing it within the realm of nature—a nature that has a history, *is* history, and engages in a form of thought distinct from that of a thinking subject or a manmade programmed machine. In short, the question Malabou's *What Should We Do with Our Brain?* poses to philosophy is: What is thinking (*Was heißt Denken?*) in light of the brain's plasticity?

Natural history ought to be contingently articulated and self-articulating. This articulation is not predetermined or programmed but rather contingent and aleatory. Nature as history cannot be conceived as continuous. Malabou's concept of plasticity serves as a tool to create space for this aleatory aspect, allowing for its expression and exploration. Brains, much like everything else, undergo self-alteration and self-formation in response to the challenges presented by the milieu. The contingency of responses is often regarded as creativity. Both the creativity of natural history and that in the human brain stem from the same source.

What about the creativity of thinking machines? Can machines also be endowed with contingency? As previously mentioned, Malabou dismisses the metaphor of a machine due to its rigidity and centralization. However, there is a moment where she embraces the notion of a plastic machine when briefly and tangentially discusses Daniel Dennett's comparison between the brain and a computer. Malabou concludes,

> What we can take away from this analysis is an approach to the machine that thinks of it not as a control center but as an organ with multiple and adaptable structures – a future-producing organization, susceptible to an always-accruing functional differentiation, a machine somehow determined by the relation to alterity – a machine capable of privileging events over laws. It is not important here to determine whether such a machine exists, but simply to insist that this conception says out loud what we live deep inside, more precisely, that "computers are not 'numbercrunching machines,'" something we experience daily, and that plasticity perhaps designates nothing but the eventlike dimension of the mechanical. (38)

Consider this merely as one example illustrating the possibility for a plastic machine. In this sense, the function of a machine extends beyond mere calculation. "Plasticity is precisely the form of our world" (38). The mechanistic image of the machine is simply an inaccurate representation. Malabou finds the concept of network to be valuable in explaining the functioning of the brain. By definition, a network lacks a central and substantial element. Networks take shape as collections of ruptures and connections, forming intricate assemblies. Furthermore, she considers the concept of a network to be pertinent to the fundamental multifunctionality of the brain (42–43). This is because each network is composed of interconnected networks, creating a system devoid of centrality or hierarchy. This is the particular way in which the tissue of the brain is constituted. The brain, as a networked tissue, does not fundamentally differ from the rest of nature in this regard, as shown in the example of Vazza and Feletti. The concept of a network serves as a mediatory tool, connecting the plasticity of the brain with its machinality. A tissue, material substance, or nature can be considered as a machine precisely because they are networked entities. Hence, what distinguishes the brain is the very quality that allows for its comparison to the entirety of nature.

Artificial Brain and Natural Machines

In *Morphing Intelligence* (2019), Malabou engages in a critical examination of her own ideas presented in *What Should We Do with Our Brain?* by exploring the intricate relationship between the "natural" brain and machines. In her 2008 book, Malabou did not regard artificial brains as authentic brains, a view that could possibly be aligned with the inherent outcome of her decentralized understanding of the brain. However, in her 2019 book, Malabou endeavors to liberate herself from many of the anthropocentric limitations. While she employs the term "natural" to refer to the human realm, and confines her discussion of machines to artificial intelligence, she nevertheless initiates thought-provoking discussions that transcend these limitations. Of particular interest is Malabou's introduction of a third category of life, distinct from both biological life and symbolic life: "the simulation of life" or what she terms "artificial life" (xvii). Malabou posits that the simulation of life is not merely an imaginary existence running parallel to real life, but instead an active participant within the realm of reality. This view adds depth to our understanding of the dynamic interplay between diverse forms of life. The concept of artificial life exemplifies the idea of life without living substance, a notion explored in Canguilhem's work in biology. It offers a compelling illustration of how life can manifest itself in forms that diverge from conventional biological frameworks. This can be observed through the fact that Malabou's reading of Darwin's studies on "domestication" and "breeding" of animals in "Whither Materialism?" (Malabou 2015, 50) reflects Canguilhem's observations in *The Normal and the Pathological*, when he considers the effect of industrial milieus identical with natural milieus (Canguilhem 1991, 143). The notion of artificial life as an authentic form of life would modify the conventional understanding of life. Moreover, the term "artificial" in this context refers to the machinic aspect. Consequently, Malabou's affirmative exploration of machinic brains in *Morphing Intelligence* can serve as a valuable conceptual tool for deepening our understanding of natural machines within the Cartesian tradition.

Within this context, we can reinterpret Malabou's acknowledgment of intelligence (the capacity for thinking) as a crucial factor that sets artificial intelligence apart from mechanistic explanations of machines and mere measurability. By emphasizing the role of intelligence, Malabou highlights the distinctive qualities and capabilities that characterize artificial life in contrast to mechanistic frameworks. In what capacity can machines be intelligent? Machines attain true intelligence when they transcend mechanical processes and are not solely manufactured based on measurable mathematical models. Instead, intelligent machines simulate life by embodying the inherent capacities for contingency and creativity, the capacities that we use to ascribe to living beings. In other words, intelligent machines must be plastic. Their plasticity entails their ability of being "subject to constant internal transformations" in response to the environment, or milieu (59). They must be living, and their brain must be truly plastic. In this context, one can draw a distinction between intelligent machines and stupid machines.[9] In line with this perspective,

[9] Take Chat-GPT, in its current form (in August 2024), as an example. While it simulates human conversations, it falls short of replicating the most intelligent human interactions. True intelligence necessitates surpassing mediocrity. But Chat-GPT lacks the capacity to transcend measurable mathematical models. A truly intelligent machine would instead simulate life, surpassing the simulation of mere mediocre (and stupid) human understanding, because life is more intelligent than human consciousness. Unlike living systems, Chat-GPT is capable to generate forms, without the capacity to produce genuine contents. In essence, a machine that merely replicates human stupidity does not merit the label of intelligence.

in the second chapter of her book, Malabou critiques the notion of IQ or the g factor as an accurate depiction of intelligence. If we adopt the understanding of life as distinct from mechanistic processes and from the inclination to reduce everything to measurement, then artificial life would indeed embody life in its authentic essence. This view invites us to reconsider and embrace the notion of artificial life as a genuine form of life, and artificial thinking as a genuine form of thinking. In her concise section on Canguilhem's 1958 work, *What is Psychology?*, Malabou encapsulates Canguilhem's critique of psychology as a discipline that is "directed against measuring and quantifying" (43). Measurement and quantification are the very aspects that diminish the nature of being human, reducing individuals to mere instruments (44). Living systems cannot be effectively studied within the confines of such sciences, and it is crucial to recognize that these systems are not positioned in opposition to non-living systems. Rather, they challenge the mechanical comprehension of systematicity and the inherent tendency towards instrumentalization within that framework. Mechanism is merely a conceptual abstraction and akin to formalism. There is no mechanism within subject *matters*. Therefore, if psychology aims to investigate genuine material intelligent brains, it would deviate from being a positivist science.

By liberating the brain from any notion of spiritual intelligence and modeling it in terms of synaptic networks, as proposed by Malabou, the essential distinction between a plastic brain and a plastic machine is effectively erased. It wipes out the differentiation between the adaptability and malleability of the brain and that of a machine. In discussing the second metamorphosis of intelligence, Malabou claims, "natural intelligence and 'synaptic' machines have the same structure" (15). She further suggests that the production of a version of computer chips which are capable of self-transformation would contribute to such reconciliation of natural and artificial brains: "The manufacture of computer chips with plasticity, that is, chips able to transform themselves" (15). These observations shed light on the pathway through which artificial intelligence can emancipate itself from strict mathematical modeling and draw closer to the realm of natural life. They underscore the potential for artificial intelligence to transcend its mechanistic origins and engage more closely with the dynamics and complexities inherent in living systems.

In their current form, our computers primarily consist of printed circuits, which renders them fundamentally non-plastic. The rigidity of these circuits limits their ability to undergo the type of transformative and adaptable processes associated with plasticity. On the other hand, the synaptic connections between nerve cells in our brain are indeed plastic, as they have the capacity for constant modifications in response to environmental changes, experiences, and learning. This plasticity allows the brain to adapt, rewire, and form new connections, enabling the dynamic nature of neural networks: "Synaptic development is therefore never the merely mechanical execution of a program or code. Rather, it depends on the synthesis between the spontaneous activity of the nervous system and interaction with the environment" (64). The disparity between computers and brains is undeniably significant. There exist fundamental distinctions that set these two systems apart in terms of their underlying mechanisms, capabilities, and modes of operation. However, Malabou, in the section "Artificial Intelligence's Counterattack", identifies "IBM's creation of an entirely new type of chip, a 'neuro- synaptic processor'" as a pivotal breakthrough in the field (83). For Malabou, this technological advancement signifies the advent of the era of "plastic" computers. This significant development in chip technology represents a paradigm shift towards computational systems that exhibit plasticity, resembling the adaptive nature of biological neural networks: "The computers of tomorrow will have processors able to adapt, self- transform, and implement their own modifications… The habitus will also affect the mode of being of the machine, which will now be capable of having an environment" (86). Malabou goes even further and identifies the future artificial machines, not only with neural networks, but with biological systems: "The future of AI is biological" (87). This view indeed signifies the ultimate identity between machines and organisms, establishing Malabou as a genuine contributor to the domain of machinic ontology: "The machines of the future will be capable of self- programming by adapting to environmental changes in real time. In this way, they will be able to modify their intensity… The characteristics of plastic machines assume specific performances, such as the evaluation of a situation, environment, or even user, in a form of mastering and self- adjustment" (90); "plasticity is not… the opposite of the machine, the determining element that stops us from equating the brain with a computer" (113). In brief, artificial intelligence has the capacity to match or even surpass natural intelligence, not by imbuing itself with consciousness or emotionality, but rather because of the machinic essence inherent in the natural brain and consciousness.

From a philosophical standpoint, this observation serves as a clear indication that machines cannot and should not be contemplated in isolation from natural entities and systems. It emphasizes the importance of acknowledging the interdependence and interconnectedness between artificial constructs and the broader natural world. Taking this perspective into account fosters a more holistic understanding of machines and their place within the intricate fabric of existence. Perhaps, it was a flawed premise from the beginning to approach the dichotomy of living systems versus machinic systems.

Is There an Anarchic History?

Malabou's recent book, *Stop Thief! Anarchism and Philosophy*, begins with the following sentences: "If Marxism is a passion for history, anarchism is the love of geography. All historical approaches, whatever their methodology, inevitably reproduce hierarchical interpretations of dominant positions" (Malabou 2023, 1). Do all historical approaches, regardless of their methodology, inevitably reproduce hierarchical interpretations of dominant power? Why is there so much emphasis on the inevitability of such collaboration? Is Malabou fundamentally opposed to history and any form of historicity? Malabou advocates for a revival of philosophical anarchism, which, according to her perspective, would be geographical and ecological. Does this signify a farewell to history? Is it possible to conceive of an anarchic history? The key point is that the anarchic geography proposed by Malabou cannot be one of borders and fixities; instead, it must be a geography of formations and deformations. The anarchic geography must be grounded in natural history. This is only possible through a radical shift in the nature of science, allowing it to incorporate contingency and natural creativity—an approach initiated in the work of Canguilhem. This implies a reconciliation between science and artistic activities, which should be at the core of philosophy. At the final stage of this chapter, we invite attention to a collection of artworks by Max Ernst, the prominent surrealist and Dadaist painter, titled "Natural History" (1926), a collection of works that is imbued with deep geological connotations, reflecting the ancestral deep history (particularly *La Mer et la pluie*, *Un coup d'oeil*, *Le Châle à fleurs givre*, and *Le Tremblement de terre*). What is intriguing about Ernst's paintings is their inhumanism—the way natural history disregards human concerns and indifferently transcends human subjectivity, while transforming it into a geological and geographical agent. Natural history petrifies human history, petrifies cities into dead forests (*La ville pétrifiée*, 1933), and this does not merely refer to past historical periods, but also to our time (*The Twentieth Century*, 1955). There is a synthesis of geology-geography and history in Ernst's work, a synthesis which is reverberated in an example that Deleuze in *Difference and Repetition* declares and implies the epigenetic heart of Malabou's thought: "The constants of one law are in turn variables of a more general law, just as the hardest rocks become soft and fluid matter on the geological scale of millions of years" (Deleuze 2014, 2). An anarchic history is that which softens geological structures and geographical borders, while anarchic geology is that which petrifies and kills human culture and values.

Therefore, the analogy between the neuronal network and the cosmic network explored by Vazza and Feletti should be reframed in light of Ernst's artistic analogy between geology and history. This reframing aims to address the ontological notion of a universe that thinks through its natural deep history. Malabou facilitates this conclusion by orchestrating connections among these realms through the concept of the plastic brain. The automatism in Ernst's Dadaism, exemplified through his technique of decalcomania, reflects the automatism of nature itself, revealing nature's creative processes, its intelligence, and the way it thinks. It is this Dadaism of natural history that Malabou calls upon for philosophizing anarchism.

References

Canguilhem, Georges. *The Normal and the Pathological.* Translated by Carolyn R. Fawcett. Zone Books, 1991.

Canguilhem, Georges. *Knowledge of Life.* Translated by Stefanos Geroulanos and Daniela Ginsburg. Fordham University Press, 2008.

Canguilhem, Georges. *A Vital Rationalist: Selected Writings from Georges Canguilhem.* Ed. Francois Delaporte. Translated by Arthur Goldhammer. Zone Books, 1994.

Chakrabarty, Dipesh. "The Climate of History: Four Theses". *Critical Inquiry* 35 (Winter 2009).

Combes, Muriel. *Gilbert Simondon and the Philosophy of the Transindividual.* Translated by Thomas LaMarre. The MIT Press, 2013.

Crutzen Paul J. and Stoermer, Eugene F. "The Anthropocene," *IGBP [International Geosphere-Biosphere Programme] Newsletter* 41, 2000.

Descartes, René. *The World and other Writings.* Translated by Stephen Gaukroger. Cambridge: Cambridge University Press, 2004.

Descartes, René. *Principles of Philosophy.* Translated by V. R. Miller and R. P. Miller. D. Reidel Publishing Company, 1982.

Deleuze, Gilles. *Difference and Repetition.* Tr. Paul Patton. Bloomsbury Academic, 2014.

Deleuze, Gilles. *Logic of Sense.* Tr. C. Boundas, M. Lester, C. Stivale. Bloomsbury Academic, 2015.

Hyppolite, Jean. *Logic and Existence.* Tr. L. Lawlor and A. Sen. State University of New York Press, 1997.

Malabou, Catherine. 2017. "The Brain of History, Or, the Mentality of the Anthropocene". *South Atlantic Quarterly* 1, 116: 39–53.

Malabou, Catherine. *Morphing Intelligence: From IQ Measurement to Artificial Brains.* Translated by Carolyne Shread. Columbia University Press. 2019.

Malabou, Catherine. *What Should We Do with Our Brain?* Translated by Sebastian Rand. Fordham University Press, 2008.

Malabou, Catherine. "Whither Materialism? Althusser/Darwin". in *Plastic Materialities: Politics, Legality, and Metamorphosis in the Work of Catherine Malabou*, eds. Brenna Bhandar and Jonathan Goldberg-Hiller. Durham, NC: Duke University Press, 2015, 47–60.

Malabou, Catherine. *Stop Thief! Anarchism and Philosophy.* Translated by Carolyn Shread. Polity, 2023.

Parsa, Mehdi. *A Reading of Gilles Deleuze's Logic of Sense.* Palgrave Macmillan, 2023.

Schuster, Joshua. "After Deconstruction? The Challenge of Malabou's Plastic Biohistory", in Wormald and Dahms. *Thinking Catherine Malabou*. Rowman and Littlefield, 2018.

Toscano, Alberto. "Plasticity, Capital, and the Dialectic". in *Plastic Materialities: Politics, Legality, and Metamorphosis in the Work of Catherine Malabou*, eds. Brenna Bhandar and Jonathan Goldberg-Hiller. Durham, NC: Duke University Press, 2015, 91–109.

Smail, Daniel Lord. *On Deep History and the Brain*. Berkeley: University of California Press, 2008.

5

Gilles Deleuze and Félix Guattari and the Abstract Machine

"Everything is a machine" (AO, 2). This quote, extracted from *Anti-Oedipus: Capitalism and Schizophrenia* by Gilles Deleuze and Félix Guattari, does not carry directly a metaphysical message about "everything" (it implies, as will be seen, a schizophrenic viewpoint). However, in this chapter, we aim to argue that Deleuze and Guattari strive to construct a distinctive metaphysical framework rooted in the concept of machine. If this holds true, then human bodies, organs, societies, as well as the organs and societies of animals and plants, planetary systems, and even the vast expanse of the universe itself, can all be machines. Following a concise examination of the concept of machine in *Anti-Oedipus*, a larger portion of this chapter is dedicated to exploring the notion of the abstract machine as expounded in *A Thousand Plateaus*. It is within this concept that a metaphysical framework, or more precisely, an ontology, takes shape.

The Desiring-Machines

The opening chapter of *Anti-Oedipus: Capitalism and Schizophrenia*, titled "Desiring-Machines", commences with the following lines:

M. Parsa, *Machinic Ontology*, https://doi.org/10.1007/978-3-031-99561-3_5

> It is at work everywhere, functioning smoothly at times, at other times in fits and starts. It breathes, it heats, it eats. It shits and fucks. What a mistake to have ever said *the* id. Everywhere *it* is machines—real ones, not figurative ones: machines driving other machines, machines being driven by other machines, with all the necessary couplings and connections. An organ-machine is plugged into an energy-source-machine: the one produces a flow that the other interrupts. The breast is a machine that produces milk, and the mouth a machine coupled to it. (1)

A few lines further, when delving into the portrayal of the schizophrenic, Deleuze and Guattari assert that "Everything is a machine. Celestial machines, the stars or rainbows in the sky, alpine machines – all of them connected to those of his body" (2). As noted, this perspective is presented from the vantage point of the schizophrenic, which, according to Deleuze and Guattari, serves as an ideal model for comprehending not only the complexities of human desire but also the intricate connections between the human body, thought, and their interplay with nature itself. A key focus of our exploration is the connection emphasized in the previously mentioned statements, which draws a parallel to the interconnectedness of machines within the schizophrenic body (the way in which machines are plugged into each other in the schizophrenic body) and the way individuals link their body-machine to celestial machines or alpine machines. In these introductory remarks in *Anti-Oedipus*, one can discern the essential attributes of machines: their capacity for coupling and establishing connections with other machines or organs, their ability to generate flows and disrupt the flows of other machines, their reciprocal drive and inter-enforcement or counter-repulsiveness, and their inherent potential for limitless expansion and extension.

At the heart of *Anti-Oedipus* lies a twofold critique: a critique of psychoanalysis that runs parallel to the critique of capitalism. Within its pages, one encounters a dialogue that encompasses a direct and unreserved critique of Freud, accompanied by a more affable critical examination of Marx. Hence, *Anti-Oedipus* revolves around discussions concerning the human psyche and the structure of human societies. Nonetheless, throughout the text, and notably in the concluding chapter, the reader will encounter ontological propositions which expand the scope of the book beyond the boundaries of human psyches and societal constructs. In this context, desire is liberated from its association solely with human bodies and is attributed to natural desiring-machines. As mentioned earlier, these references in *Anti-Oedipus* may appear scattered. However, it is in the second volume of *Capitalism and Schizophrenia* that Deleuze and Guattari delve more explicitly into the logic, ontology, and metaphysics of machines.

Anti-Oedipus primarily explores the nature of desire, which, from a psychoanalytical perspective, is inherently unconscious desire. Desire, in its essence, does not emanate from an individual or subject. Rather, a person or subject may *need* something that already exists. However, desire itself is fundamentally distinct. Desire is far more primal, serving as the impetus or driving force that propels the functioning of organs and systems within an intricate network, enabling them to operate and engage with other organs and systems. The question is: why does the heart beat and the lungs breathe? In classical and religious frameworks, the answer would suggest they serve a higher authority, namely the commanding organ (the brain), guided by a specific telos. Why do birds and humans migrate? Because the holy nature or God command them. Because everything is planned by the Lord. According to Deleuze and Guattari, psychoanalysis, much like established religions, tends to focus on unconscious desire only after it has been subjugated and confined within familial structures and systems of power. In doing so, psychoanalysis unifies desire, personalizes it, and subjects it to the predetermined role of the father. At this juncture, the critique of Freud becomes entwined with a Marxist interpretation of history. Various mechanisms of despotism or capitalism employed within the structures of social power have the effect of diminishing the untamed nature of desiring machines, transforming them into compliant and docile individuals. These power structures impose constraints and norms that suppress the inherent potency and creativity of desires. The Oedipal family assumes a pivotal role in this context, as it serves as the conduit through which the power of the despot is transmitted to individuals. Specifically, the centralized authority of the despot is channeled through the father figure, allowing for the individualization and control of corporeal forces and desires. This mechanism reinforces the power dynamics within society and establishes a framework for the regulation and normalization of desires. Deleuze and Guattari embark on a quest to uncover desires that precede their subjugation within oedipal triangles. In the case of the schizophrenic, they explore a notion of desire which remain unbounded by oedipalization, liberated from paternal domination and the negative desire directed towards the mother. By doing so, they challenge the social constructs rooted in despotic or capitalist frameworks that are built upon notions of needs and interdependence. Their exploration seeks to reveal a notion of desire that defies these restrictive structures, fostering a space for liberated and autonomous expressions of desire. Thus, alongside their exploration of the unconscious desires of the schizophrenic, Deleuze and Guattari also strive to envision a society free from the dominance of State apparatus in its various historical manifestations and modes of governance. Simultaneously, they seek to conceive of a nature emancipated from the constraints imposed by the laws of nature. Free desire, free society, free nature. Schizophrenic desire versus sovereignty and state science. Consequently, desire becomes a concept encompassing the liberated driving forces that propel everything in motion, everything that exists. The reason why the figure of the schizophrenic assumes a crucial role in Deleuze and Guattari's critique of psychoanalysis is that it embodies a desire that exists beyond the confines of oedipal closure, surpassing the limitations imposed by the constructs of family, state, and societal laws. The schizophrenic becomes a symbol of liberation, embodying a desire that transcends conventional boundaries and offers an alternative perspective on the complexities of human existence. The schizophrenic is wild, uncultivated, and psychoanalysis is too weak to analyze nature in its purity. This is where schizoanalysis is invited to step in.

According to Deleuze and Guattari, a fundamental distinction between unconscious desire and conscious need lies in the fact that desire produces its own objects. The objects of desire are not preexisting entities awaiting consumption. This characteristic is pivotal in understanding desire as machinic. Desiring-machines, by their very nature, serve as generators. In this sense, the unconscious can be understood as a machine, deviating from Jacques Lacan's perspective that it is solely a structure or structural language. Rather than being confined to linguistic or structural frameworks, the unconscious operates as a machine, generating processes and connections, as well as affects and emotions. As already mentioned, considering the unconscious as real machine not only de-oedipalizes the unconscious, but also de-personalizes it. By conceptualizing the unconscious as a machine, it shifts the focus away from individual subjectivity and personal narratives. There are two avenues through which this depersonalization occurs, signifying two distinct paths for extending the unconscious beyond individual boundaries: the social machine and the natural machine. The unconscious serves as the site of production for both social objects and natural objects. Within the realm of the social, the unconscious generates and shapes social constructs, institutions, ideologies, and cultural phenomena that define our collective existence. It is through the workings of the unconscious that social objects, such as languages, norms, customs, and symbolic systems, come into being and exert their influence. This domain aligns with aspects of structuralism, as espoused by Jacques Lacan, who places the symbolic order at the core of understanding unconscious desire. It also resonates with the concept of social constructivism, which posits that all objects are social objects. Deleuze and Guattari, in *Anti-Oedipus*, lean towards exploring the alternative direction of how unconscious desire produces natural objects, while also recognizing the intertwining of social and natural production. But how can the unconscious produce the natural objects? How the social unconscious is compatible with a natural unconscious?

The second part of the fourth chapter of *Anti-Oedipus*, titled "Molecular Unconscious", offers a comprehensive explanation of this concept. The distinction between the unconscious and consciousness is compatible with the differentiation between the molecular body without organs, which serves as the intense foundation for any process of individuation, and the molar individualities. Therefore, following the incorporation of this physical differentiation between the molar and the molecular into their understanding of the unconscious, Deleuze and Guattari assert,

> … in reality the unconscious belongs to the realm of physics; the body without organs and its intensities are not metaphors, but matter itself. Nor is it our intention to revive the question of an individual psychology and a collective psychology, and of the priority of the one or the other; this distinction, as it appears in *Group Psychology and the Analysis of the Ego*,

> remains completely stymied by Oedipus. In the unconscious there are only populations, groups, and machines. When we posit in one case an involuntariness *(un involontaire)* of the social and technical machines, in the other case an unconscious of the desiring-machines, it is a question of a necessary relationship between inextricably linked forces. (AO, 283)

Deleuze and Guattari use the term "unconscious" to refer to something distinct from both the individual and collective unconscious, which are concepts discussed by Freud in *Group Psychology and the Analysis of the Ego*. The crucial point is that the expansion of the unconscious into the social realm does not free it from the grip of the Oedipus. The unconscious should be completely transferred to the realm of physical nature. The primary objective behind employing the term "molecular unconscious" is to emphasize the complete integration of the unconscious with the domain of physical nature. Only under these circumstances, where we shift our focus from the physical unconscious back to the psychic and social unconscious, can psychoanalysis (renamed now as schizoanalysis) offer a real and non-Oedipal understanding of our desire.

Deleuze and Guattari emphasize that the molecular unconscious corresponds to the unconscious of the desiring-machines. Therefore, desiring-machines are essentially molecular machines. But what exactly are these molecular machines? They refer to a physical domain that, on one hand, stands in opposition to mechanism and, on the other hand, opposes a vitalist organism:

> From machines, mechanism abstracts a *structural unity* in terms of which it explains the functioning of the organism. Vitalism invokes an *individual and specific unity* of the living, which every machine presupposes insofar as it is subordinate to organic continuance, and insofar as it extends the latter's autonomous formations on the outside. (AO, 284)

The conclusion drawn is that machines lack any form of unity, be it structural or organic. Machines, made up of restless drives and desires, are characterized by the fundamental movement of *deterritorialization*, a concept elaborated in the second volume of *Capitalism and Schizophrenia*, *A Thousand Plateaus*. The functioning of a machine deterritorializes its unity and identity, drawing it outside of itself. This depersonalizes the machine, enabling the eventual liberation of the concept of the machine from its societal implications in *Anti-Oedipus*, and ultimately leading to the idea of nature (the Earth) as a mega-machine in *A Thousand Plateaus*. In this sense, the transition from *Anti-Oedipus* to *A Thousand Plateaus* marks a shift from a schizophrenic-based reading of the machine-thesis ("everything is a machine") to an ontological perspective that encompasses nature in its entirety. This represents a radicalization of the approach taken in the previous chapter on Catherine Malabou: moving from the plasticity of the human brain to natural history—the Earth—as a plastic brain.

Moreover, the molecular machine transcends the differentiation between mechanical objects and vital organisms, as both of them belong to the realm of molar (conscious) entities. Classical phenomenology, which is concerned with the study of consciousness, only has access to these mechanical objects and vital organisms. Any intended unity, whether structural or organic, belongs to the realm of the molar. Molecular machines are inherently composed of irreducible multiplicities. No molecular machine can be reduced to a single unified entity. Machines are not things or objects. Both mechanical machines (such as systems of stones, planets, etc., including manmade machines) and organic systems are composed of molecular machines; "a domain of nondifference between the microphysical and the biological, there being as many living beings in the machine as there are machines in the living" (AO, 286).

Similarly, machinic production differs from both mechanical production and organic reproduction, which are secondary in relation to the primary process of molecular machinic production. Drawing upon Samuel Butler, Deleuze and Guattari characterize machinic production as the "surplus value of code", "when a part of a machine captures within its own code a code fragment of another machine, and thus owes its reproduction to a part of another machine" (AO, 285). In this mode of production, which is the only genuine form of production, the form of the produced entity becomes apparent only after the process is developed. As a result, it can never be premeditated or consciously intended. This is precisely why machinic production can only occur through the agency of the molecular unconscious, which operates at the level of nature.

From Desiring Machines to Abstract Machines

A significant theme explored in *A Thousand Plateaus* is the transformative journey of unformed matter as it assumes form, giving rise to perceivable objects and entities. Drawing upon Simondon's thought, Deleuze and Guattari acknowledge the priority of a pre-individual state and strive to formulate its operations within the framework of the logic of pure multiplicities. Furthermore, in alignment with Simondon's ideas, they present a scientific analysis of this state that transcends conventional notions of tangible objects and instead relies on scientific abstraction. This involves the introduction of molecular machines that operate at the fundamental level of nature, giving rise to the formation of molar machines.

In *A Thousand Plateaus*, machines are the elements of a logic and ontology of multiplicities which are indeed, "acentered systems, finite networks of automata in which communication runs from any neighbor to any other, the stems or channels do not preexist, and all individuals are interchangeable" (ATP, 19). Here, the concept of "machinic multiplicity" stands in contrast to centralizing, unifying, or organic multiplicity. The latter is derived from the former through a process of double articulation known as "stratification", which serves to generate diverse objects. Deleuze and Guattari extensively draw upon scientific discoveries in molecular biology and geology to discuss the emergence of mechanical objects and organic systems from the realm of pure molecular multiplicities or bodies without organs. These refer to the unformed matter that has yet to be structured into relatively stable objects or organisms. In the initial Plateau of *A Thousand Plateaus*, Deleuze and Guattari employ the concepts of tree and root on one hand, and rhizome on the other, to differentiate between these two types of multiplicities. The former represents a structure that operates under the dominance of a transcendent principle, leading to its self-contained nature. In contrast, the latter embodies a machine that remains open, capable of plugging into and extending onto other machines and multiplicities. The process of plugging does not occur between fully-formed machines, but rather, it generates and produces them. Hence, there exists a distinction between molecular or pure machines and molar machines. Pure machines are inherently rhizomatic, meaning they lack a radicle or central commanding center, and they disregard any notion of interiority. Pure machines are characterized as open machines.

Based on this premise, Deleuze and Guattari define the process of abstraction from dominant power and its products as the intrinsic machinic movement, expressed through the concepts of the line of flight or deterritorialization. This is not merely a distinction between two types of machines, one closed and the other open. Instead, the notion of a machine entails a transformative movement that dismantles the self-contained nature of molar machines, or as Deleuze and Guattari describe it, deterritorializes them. This is the essence of the term "abstract" in the context of abstract machines. The machinic movement encompasses the process of abstraction away from consciousness and its creations, the state apparatus, and any rigid structures: "Multiplicities are defined by the outside: by the abstract line, the line of flight or deterritorialization according to which they change in nature and connect with other multiplicities" (ATP, 9–10). A territory, shaped by the exercise of power, embodies an identity or a type. The line of flight or the movement of deterritorialization is a revolutionary change or a phase-change that alters the type and changes the nature of the identity. In this way, an organism metamorphoses into a monster, an object transforms into a (natural) artwork. This movement is far from peaceful, as it challenges the established power and disrupts the hierarchy of its subordinates and the essence of territories. Any genuine machine contains the lines of abstraction from territories, making it an abstract machine. Every true unconscious is a political unconscious. Given the inherently non-peaceful nature of these movements, abstract machines can be regarded as war machines, functioning in accordance with the principles of "guerrilla logic" (19). This signifies a war not only against social domination but also against the psychoanalytic domestication of desire. Moreover, as will be elaborated, this war goes beyond serving or occurring between state apparatuses or any other power structures, but is based mostly on the idea of a total civil war.[1] Or more precisely, the war operates at the molecular level.

---

[1] Eric Alliez and Maurizio Lazzarato, in *Wars and Capital* (2016), build a theory of global or generalized civil war, which is continued, with other means, through economy. Their theory is based on,

The concept of the machine aids in dispelling any centralized and authoritative perception we might hold regarding our bodies, societies, and nature. The notion that everything is a machine implies the absence of a stable substance. However, it does not denote a coherent and simplistic flow that can be effortlessly assimilated by other flows. Rather, it represents a complex structure that can interconnect with other structures. For instance, the brain does not function as a commanding center of our body, as it is inherently machinic. The brain can be understood as a molecular machine in the sense that it lacks an arborescent structure and instead exhibits a rhizomatic organization, devoid of hierarchical arrangements; "… the brain is not a rooted or ramified matter… Many people have a tree growing in their heads, but the brain itself is much more a grass than a tree" (ATP, 17). As mentioned earlier, this is a radicalization of Catherine Malabou's concept of the plastic brain, in which she also attempted to apply this plasticity to artificial brains, as discussed in the previous chapter. Hence, Deleuze and Guattari's assertion applies to artificial brains as well, or more specifically, they advocate for the concept of rhizomatic AI:

> Arborescent systems are hierarchical systems which centers of signifiance and subjectification, central automata like organized memories. In the corresponding models, an element only receives information from a higher unit, and only receives a subjective affection along preestablished paths. This is evident in current problems in information science and computer science, which still cling to the oldest modes of thought in that they grant all power to memory or central organ. (ATP, 18)

Consequently, we can discern two types of information machines by considering two contrasting images of thought and the brain, as well as two distinct conceptions of automatism. One image portrays a centralized and arborescent structure, while the other embodies a rhizomatic and molecular configuration. As the examples of this automatism, one can consider the nervous system and the social system. Interestingly for our purpose, Deleuze and Guattari take often both

---

among others, Deleuze and Guattari's concept of war machine. Their theory will be discussed briefly in the final section of this chapter.

systems in tandem, insisting that in both cases a molecular automatism is in function. In the ninths Plateau of *A Thousand Plateaus*, "Micropolitics and Segmentarity", while discussing the segmentarity of various social systems, which is another name for their molecularity, Deleuze and Guattari write, "Perhaps the opposition sociologists establish between the segmentarity and the central is biological deep down: the ringed worm, and the central nervous system. But the central brain itself is a worm, even more segmented than the others, in spite of and including all of its vicarious actions. There is no opposition between the central and the segmentary" (231). In this sense, the molecular machinic unconscious can be exemplified in the monstrous artificial intelligence that can operate as a guerrilla war machine challenging the bureaucratic or administrative machines of our rational social systems. It is solely for the purpose of safeguarding our rigid bureaucracies that we incessantly propagate our apprehensions of soulless intelligence.

Machines operate and generate various effects; however, they are never monofunctional. No machine is a unified being. This aligns with the concept of the interior milieu in Canguilhem and Simondon, which acknowledges the inherent contingency of any technical or organic object. This implies that the logic of machines resides in productive multiplicities, rather than consummative unities. The multitude of organs and functions cannot be reduced to a singular unity or identity, such as a brain, personality, or people. When discussing the unconscious-machine, Deleuze and Guattari assert, "One of the essential characteristics of the dream of multiplicity is that each element ceaselessly varies and alters its distance in relation to the others… These variable distances … cannot increase or diminish *without their elements changing in nature*" (34, emphasis in original). This concept has been previously addressed in our preceding chapters: real and natural machines are inherently non-mechanical, which applies to both natural and artificial brains, as well as other systems. Machines exhibit boundless multifunctionality and generate unprogrammed, unseen objects due to the perpetual variation of their elements and the constant alteration of distances between them. Once again, Deleuze and Guattari employ a physical concept to describe the elements of machines: "their movements are Brownian" (37). At a micro or sub-molecular level, the Brownian movements exemplify the non-mechanical nature of physics. According to Simondon, it is precisely this aspect of physics that serves as the paradigm of individuation, permeating all that exists. In this context, micro-machines, regardless of their size (even the Earth, which Deleuze and Guattari regard as a "giant molecule"), possess the primary characteristic traditionally attributed to living beings: the unpredictable movements of their elements, or simply put, freedom—the unconscious freedom. In this sense, the unconscious is inherently libidinal, and this holds true for both the social and natural unconscious. All true machines can be considered bio-machines in this sense. Deleuze and Guattari also refer to this aspect in terms of molecularity. Molar machines consist of molecular machines, and this distinction is not based on size but rather on their behavior and mode of operation. The mechanical machines, whose operational laws can be formulated by natural sciences, are comprised of molecular machines that necessitate a distinct kind of natural science capable of generating concrete and contingent rules. Similar to Canguilhem and Simondon, Deleuze and Guattari employ the concept of non-organic life in this context: "the libido suffuses everything … a social machine or an organized mass has a molecular unconscious that marks not only its tendency to decompose but also the current components of its very operation and organization" (ATP, 39).

This represents an ontology, or more precisely, a metaphysics of machines as it seeks to comprehend the diverse levels of reality. It can even be extended to encompass the entirety of reality as a whole: "We can no longer even speak of distinct machines, only of types of interpenetrating multiplicities that at any given moment form a single machinic assemblage, the faceless figure of the libido" (ATP, 41). What we perceive as individual machines are indeed molar machines, but at the molecular level, there exists only a single (infinitely manifold) machine, the mega-machine, encompassing nature itself with its inherently "wild multiplicities" (ATP, 42). One can discern Deleuze's thesis of the univocity of being within the pages of *A Thousand Plateaus*. Hence, the statement "There has only ever been one ontological proposition: Being is univocal" (DR, 43) finds its corresponding counterpart in "Flat multiplicities of n dimensions are asignifying and asubjective" (ATP, 10). Or, "Philosophy merges with ontology, but ontology merges with the univocity of Being … The univocity of Being does not mean that there is one and the same Being; on the contrary, beings are multiple and different" (LoS, 179), can be traced in *A Thousand Plateaus* as "… to move between things, establish a logic of the AND, overthrow ontology, do away with foundations, nullify endings and beginnings" (28). *A Thousand Plateaus* translates the univocity of being in terms of the machinic nature of multiplicities, which does not imply a multiplicity of beings but rather pure and wild multiplicities that cannot be encompassed or domesticated within an ontology of beings. It entails therefore a wild machinic ontology. It is only under these circumstances that one can expand the realm of machinic multiplicities to encompass a universal yet non-unifiable domain. This essentially implies that what applies to local machines also applies to the mega-machine; "… the same machine at work in astrophysics and in microphysics, in the natural and in the artificial …" (ATP, 62).

Machines have bodies and forms. Pure or abstract machines are formless. The body of a machine can be understood as the body without organs. Abstract machines consist of unformed matter, free matter, or what we previously referred to as chaos in our examination of Simondon. As we explored in our chapter on Simondon, chaos is not only unformed but also informed, as it is replete with potentials for generating forms. "Unformed matter, the phylum, is not dead, brute, homogeneous matter, but a matter-movement bearing singularities or haecceities, qualities, and even operations (itinerant technological lineages)" (ATP, 563–564). In a particular sense, nature can be considered living, as it is imbued with a primal libido. It is living without being organized:

> the body without organs is opposed less to organs as such than to the organization of the organs insofar as it composes an organism. The body without organs is not a dead body but a living body all the more alive and teeming once it has blown apart the organism and its organization. (ATP, 34)

Both the concepts of machine and life stand in opposition to that of a structured organism. Machines are infused with non-organic life. In the concluding Plateau, Deleuze and Guattari write, "there is no reason to think that all matter is confined to the physicochemical strata: there exists a submolecular, unformed Matter. Similarly, not all Life is confined to the organic strata: rather, the organism is that which life sets against itself in order to limit itself, and there is a life all the more intense, all the more powerful for being anorganic. There are also nonhuman Becomings of human beings that overspill the anthropomorphic strata in all directions" (554). Does this inclination towards the inhuman also signify a shift towards encompassing artificial life as an integral component of life itself? Is the destiny of life on Earth destined to be shaped by the triumph of machines over human consciousness, and artificial life over organic life?

Abstract Versus Mechanical

In *A Thousand Plateaus*, a distinction is made between two concepts of machines: molar and molecular, customary and abstract. On one hand, there are the machines that organize and territorialize, while on the other hand, there are the machines that mark the lines of flight and facilitate the movement of deterritorialization. We have also referred to them as closed and open machines. However, this distinction is not the same as the difference between abstract machines and mechanical machines. In Deleuze and Guattari's ontology, customary mechanical machines, when considered within the context of their historicity, natural constitution, and the various layers of their functioning, are not as monofunctional as they are commonly assumed to be. While they are undoubtedly posterior in relation to absolute abstract machines, customary mechanical machines still adhere to the same concrete rules. Machines are distinct from mere tools. All machines possess a molecular automaticity.

This should serve as the focal point of our understanding regarding the essence of technics and artifacts. At the individual molar level, machines function in accordance with the predefined tasks we have consciously assigned to them. However, when we consider them at a historical or collective level, machines no longer adhere to these predefined definitions. In his celebrated text "Machine and Structure", Guattari defines "the history of technology" as follows: "Every machine is the negation, the destroyer by incorporation (almost to the point of excretion), of the machine it replaces" (Guattari 1984, 112). According to Guattari, historical analysis transforms every structure into a machine by replacing the structural relations between theories—a totalitarian and ideological interpretation of the history of science—with machinic, destructive relations between structures.

What does this tell us about the essence of technology? As discussed by Heidegger in *Question Concerning Technology*, the essence of technicity is far removed from instrumentality: "the correct instrumental definition of technology still does not show us technology's essence" (Heidegger 1977, 6). Heidegger makes a compelling effort to establish a connection between the essence of technology and the poietic act of nature itself, emphasizing that our relationship with it should be understood in terms of *Aletheia* and truth. He distinguishes the essence of technology from the modern goal of exploiting nature and endeavors to establish a connection between this essence and the artistic force that binds us to the creative essence of nature. Deleuze and Guattari acknowledge and respect the Heideggerian insight that distinguishes the essence of technology from the mere substantiation that arises from instrumentalization. However, Deleuze and Guattari's perspective differs from Heidegger's by emphasizing the separation of the creative essence of nature and technology from the phenomenological notions of revealing and truth.

Machines lack any identity or substance. When considering the concept of the abstract machine, its abstraction should also be understood in relation to forms and substances. Machines are abstract in relation to the platonic abstract Ideas and forms. In this regard, the abstraction of machines is a movement towards the concreteness of the real, akin to the approach presented in Gaston Bachelard's *The Formation of Scientific Mind* in which the abstractness of science (what he calls "truly healthy, truly dynamic scientific abstraction"; Bachelard 2002, 64) aids in disengaging from the ordinary mode of thinking which is trapped within itself and has no access to the real as such.[2] In both cases—the formation in the educational sense as intended by Bachelard and the pragmatism of *A Thousand Plateaus*—it is crucial to constantly bear in mind that reality is

[2] Bachelard, in this regard, writes, "Our intention in this book is to show this grand and imposing destiny of abstract scientific thought. In order to do so, we must prove that abstract thought is not synonymous with a bad scientific conscience, as the common accusation seems to imply. We shall have to prove that abstraction clears the mind of encumbrances, that it lightens the mind and makes it more dynamic" (Bachelard 2002, 18).

not a stable structure to be merely recognized or represented, but should be viewed as something to be produced and to be operated. In this sense, Deleuze and Guattari assert, "Abstract machines operate within concrete assemblages" (ATP, 562). Similar to Bachelard and Canguilhem's views to scientific thought, this aligns with our understanding of the scientific approach to the real. If nature is endowed with non-organic life, then the science of nature should embrace a non-mechanical micro-science which is equipped with a dynamic healthy abstraction. Once again, it is important to emphasize that science is not merely a tool for representation, but rather an active and transformative intervention in shaping the formation of the real. Technology is not merely a byproduct of the scientific approach; rather, it is an integral and inherent component of science itself. For Deleuze and Guattari, what is at stake is not only the relationship between the laws of nature and the abstraction of machines, but also the interplay between the concreteness of rules (ontological pragmatism) and the abstract nature of machines. Hence, they maintain,

> … abstract machines know nothing of forms and substances. This is what makes them abstract, and also defines the concept of the machine in the strict sense. They surpass any kind of mechanics. They are opposed to the abstract in the ordinary sense. Abstract machines consist of unformed matters and nonformal functions. Every abstract machine is a consolidated aggregate of matters-functions (*phylum* and *diagram*). This is evident on a technological "plane": such a plane is not made up simply of formed substances (aluminum, plastic, electric wire, etc.) or organizing forms (program, prototypes, etc.), but of a composite of unformed matters exhibiting only degrees of intensity (resistance, conductivity, heating, stretching, speed or delay, induction, transduction…) and diagrammatic functions exhibiting only differential equations or, more generally, "tensors". Of course, within the dimension of the assemblage, the abstract machine, or machines, is effectuated in forms and substances, in varying states of freedom. (562)

Machines are considered abstract in the sense that they lack definite forms and substance. However, this is not a description of the existence of machines, but rather an explanation of their (nonformal) functionality. Machines produce and their very act of production alters both their own nature and the nature of the milieu in which they operate. Machinic production differs significantly from mechanical production. In the case of the latter, the production process does not bring about any changes to the configuration of the mechanism. Machinic production can be likened to how a change within an ecosystem leads to the emergence of a new species, which subsequently influences and modifies the ecosystem itself. Or, on a technological plane, the composition of unformed materials and diagrammatic functions gives rise to the creation of new machinic assemblages that have their own modes of agency and autonomy. Therefore, the term "abstract" can also encompass the notions of monstrosity and the unseen. By introducing abstract machines, Deleuze and Guattari assert the existence of monstrous machines, but in a broader sense that implies non-mechanical information-monster-machines (GAI) as well as non-organized natural bio-machines (a tortoise, a pack, a colony). The motor of abstract machines in this sense is the irreducibility of multiplicities (52–53). In this context, the concept of Bachelardian healthy abstraction serves as the sole means to access Canguilhem's domain of the pathological. Hence, the healthy sick machines are the disjunctive syntheses at the broken heart of *A Thousand Plateaus*. This synthesis occurs beyond human consciousness, within the realm of the machinic unconscious. A prime example of Deleuze and Guattari's machinic unconscious would be a monstrous artificial intelligence.

In this manner, the epistemology of Bachelard and Canguilhem aligns with the ontology of machines. Therefore, we can rephrase the characteristics of the "minor science" as described in the epistemological "Proposition III" found in Plateau twelve of *A Thousand Plateaus*, when considering abstract machines (ATP, 398–399). According to Deleuze and Guattari, the concept of minor or nomad science, contrasting with royal or State science, mirrors Canguilhem's approach to the life sciences as opposed to mechanical sciences. In this context, we can summarize the defining characteristic of abstract machines as being the subjects of nomad sciences: 1, abstract machines are hydraulic machines, which is to say, they are inseparable from flows, and no theory of solids is applicable to them. The flows are constituents of individuals, while at the same time, they are inherently social. 2, the constitutive flows of machines are heterogeneous, rendering the machines in a constant state of becoming. This inherent characteristic makes it impossible to classify the abstract machine as a fixed entity or substantial object. 3, machines possess a vortical nature, which serves as their distinguishing characteristic. It is through this vortical quality that flows generate distinct machines, even though, from a cosmic perspective, they can all be regarded as components of the same mega-machine. At the core of every abstract machine lies a turbulence, fueling its essence. 4, abstract machines are affective. They cannot be recognized or known. Instead, they manifest themselves as events, devoid of describable effects, but rather conveying expressible affects.

Political Metallurgy

The distinction between molecular and molar machines, as well as their order of priority, aligns with the differentiation of two types of war machines: one being molar and the other molecular. Of course, the true war machine is molecular, marking the real movement of abstraction from territories. However, there always exists a force that seeks to territorialize the war itself within the confines of national and State identities. Hence, the wars between the proprietors of territories, namely nations or states. While deterritorialization may indeed occur during State wars, it would be incorrect to ascribe the role of causality to molar wars. According to Deleuze and Guattari, "following Clausewitz's intuition, the war machine is very different; it is a flow of absolute war stretching between an offensive and a defensive pole, and is marked only by quanta (psychic and material forces that are like the nominal liquid assets of war)" (ATP, 240).

This brings us to a pivotal differentiation between two distinct notions of politics, which are implicitly delineated throughout the course of *A Thousand Plateaus*: micro-politics and macro-politics. Just as Deleuze and Guattari broaden the term "machine" beyond its conventional definition as a mechanical device to encompass an ontological realm, they also subject the terms "war" and "politics" to a similar transformative journey in their work. The transition unfolds from molar war and molar politics to the realm of molecular war and molecular politics. Under this circumstance alone can one assert that everything is political or perceive politics as the continuation of wars with other means: "everything is political, but every politics is simultaneously a *macropolitics* and a *micropolitics*" (ATP, 235).

In the twelfth Plateau of *A Thousand Plateaus*, Deleuze and Guattari elaborate more on the concept of molecular war, which transcends the notion of being either for or against a State apparatus, and instead aligns with psychic and material forces. Drawing upon Heraclitus and Nietzsche, Deleuze and Guattari elevate the concept of war to an ethico-ontological plane, perceiving it as an operation aimed at radical transformation of the mode of social life. They define the operationality of war machine in opposition to the functionality of the State apparatus: "As for the war machine in itself, it seems to be irreducible to the State apparatus, to be outside its sovereignty and prior to its laws" (388); "In every respect, the war machine is of another species, another nature, another origin than the State apparatus" (389). Evidently, the concept of machine in general is implicitly contrasted with that of the apparatus. The State apparatus functions by means of fixation and control, internalization and preservation, employing prisons, police officers, educational systems, families, and various other mechanisms to conserve its established state of being. However, according to Deleuze and Guattari, this state of affairs is not an inevitable outcome of economic development within primitive societies. It is inherently counterproductive in nature. "The concern of the State is to conserve" (394). In contrast, machines are inherently hydraulic, driven by the force towards deterritorialization. In essence, all machines can be considered war machines, aligning with the ontological understanding of war as articulated in the thoughts of Heraclitus and Nietzsche. All machines are quantum machines, composed of "quantum flows (micropolitics) that continually reshuffle and stirs up" their "segments" (240). From this perspective, *A Thousand Plateaus* can be seen as an extension of *Anti-Oedipus*, where the critique of the Oedipal family in the latter work is extrapolated to the broader social-political realm in the former.

Deleuze and Guattari reject the notion that the State emerges at a specific moment in the course of social development. They argue that the State has perpetually existed as a counterproductive force, seeking to preserve the existing state of society, regardless of whether it is primitive, barbaric, or modern. The distinction between primitive and modern societies lies not in the former being decentered and the latter being centralized under a State apparatus, but rather in the nature of segmentarity or social structuration. In primitive societies, segmentarity is supple and plastic, whereas in modern societies, it becomes rigid and forms an arborescent system, although Deleuze and Guattari insist that in either society both models exist, overlap and are entangled (ATP, 231–234). However, the State apparatus can be considered conceptually as seeking the mechanization of the social machine, which, similar to any domain of nature, possesses an inherently productive and generative nature. As a result, the State war becomes a negative instrument employed in the preservation of the State, rather than the civil war or a positive force that drives deterritorialization: "*The state has no war machine of its own*; it can only appropriate one in the form of a military institution, one that will continually cause it problems. This explains the mistrust States have toward their military institutions, in that the military institution inherits an extrinsic war machine" (391). Interestingly, Deleuze and Guattari position the State and the essence of war in opposition to one another, asserting that no army can be completely instrumentalized to serve the interests of the State, regardless of how the State portrays its military as a symbol of absolute subjugation to its laws. In this context, we can examine the relationship between revolutions and coup d'états. Both originate from rhizomatic roots outside the confines of the State apparatus, with revolutions stemming from the molecular machine and coup d'états emerging from the realm of molecular war. Within the notion of war machine, revolutions and coup d'états are identical, dividable into molecular and molar registers.

No totality can totalize the nature, no State apparatus can make a perfect interiority. This is why wars are inevitable. However, the machinic ontology proposes a perverse totalization of multiplicities within a megamachine that resonates with the idea of the communication of all events in the Event in Deleuze's *Logic of Sense*.[3] This represents Deleuze and

---

[3] In *Logic of Sense*, the communication of events presents an alternative to the traditional logic and ontology of essences and substances, introducing the Event as an ethical-logical universal. In this perspective, events are inherently disjunctive, and the only way to establish communication among them is through a resonance of heterogeneous series. The differentiation between the logic of essences and the logic of events (or sense) serves as an early foundation for distinguishing between

Guattari's interpretation or reformulation of the communist slogan found in Marx and Engels' 1848 Manifesto: "Workers of the world, unite!" The unity of multiplicities, which is indeed a univocity, manifests itself across all domains that exist beyond the reach of State apparatuses, "huge world-wide machines … but also the local mechanisms of bands, margins, minorities, which continue to affirm the rights of segmentary societies in opposition to the organs of State power" (ATP, 397). The unity of workers (proletariats) is not an organic unity, but rather a machinic (allagmatic) one. They are actors in networks, not organs in organisms. At the onto-political level, which can also be referred to as libidinal politics, this encompasses the geological, biological, and, ultimately, the ontological ground upon which States and social groups are surfaced. In this context, communism as the condition for the possibility of alternatives entails a crucial distinction between the organic unity enforced by the State and the machinic unity that derives its strength from the revolutionary impulses of war machines.

Deleuze and Guattari describe the social machine, which is not striated within a State apparatus, in terms of nomads characterized by their ability to move and freely distribute themselves within smooth spaces, such as deserts. The nomad is a society that lacks a substantial interiority and can be defined based on the smooth space of its distribution, which serves as the milieu for its formation. The nomadic trajectory "*distributes people (or animals) in an open space*" (420). This is a consequence of social machines being composed of various flows and the syntheses that occur among them. However, it becomes evident that the flows within social machines are not solely limited to liquid flows.[4]

the ontology of beings and that of multiplicities in *A Thousand Plateaus*. Refer to the Twenty-Fourth Series of the Communication of Events in *Logic of Sense* for further exploration (LoS, 175–182).

[4] In Werner Herzog's *The Enigma of Kaspar Hauser* (1974), Kaspar Hauser becomes a poignant representation of the nomad's dream, portrayed as a vast desert through which the caravan of human existence is guided by a blind ruler. The film's emphasis on the unknown rest of the story underscores the contingent nature of nomadic distribution. Furthermore, the movie brilliantly explores the juxtaposition between nomad science and royal science, as discussed in the Twelfth Plateau of *A Thousand Plateaus*, along with the distinction between a logic of sense and the logic of essences.

The crucial aspect lies not in the distribution of the nomad, but rather in the nomadic distribution itself, which signifies the manner in which it operates on the earth's mega-machine to generate customary machines. The emphasis lies not on the act of distributing within a space, but rather on the distribution of the space itself. The earth is the territory of the nomad and of the nomadic distribution. The earth is the body without organs from which all entities, organisms, and organizations emerge. The concept of nomadic distribution encompasses both natural history and social-political history, emphasizing the intrinsic connection between the two. The distribution of societies and political power on Earth corresponds to the distribution of the Earth itself, forming an inseparable connection between these societies and the material forces and flows of the Earth, including the technical configuration of its layers and tectonic plates.

What does it have to do with the history of human technicity? It serves as an intermediary phase between the nomadic distribution of the Earth and the political historicity of human life on the Earth. On this basis, Deleuze and Guattari distinguish between a royal history and a nomad history. The royal history isolates man and its historicity, whereas the nomad history regards it as an integral part of the Earth. Furthermore, concerning human technicity, in the royal reading of history, which is the history of State apparatuses and their equivalents, man invents tools, while in the nomadic reading, man invents weapons; utilitarian agricultural machines versus war machines that turn hunting into war; work versus free action; culture versus nature. In this way, Deleuze and Guattari trace the source of technicity in the free and creative element of nature, in the natural historicity and the fundamental war at the heart of nature that necessitates the invention of weapons.

Therefore, Deleuze and Guattari associate the nomad history with the age of metal. A fundamental distinction arises between soil and metal in terms of history, just as there is a distinction between blood and flesh. On this basis, in the Twelfth Plateau of *A Thousand Plateaus* titled "Treatise on Nomadology - The War Machine", Deleuze and Guattari discuss the role of metallurgy. Crucible steel is that which connects the earth and its organs, bridging the realms of nature and culture. The vast expanse in which nomads traverse empowers them to effortlessly mold the malleable substance of metal using the strength of their own flesh. This does not imply territories, possessions, or heredities. Nomad's metallurgy is a participation in a process that already exists within nature. Metallurgy is the common birth-point of art and science. It is a nomad science and nomad art that embraces boundless variations, where the act of formation and deformation intertwine as a unified process: forging as the meeting point of the artificial and the natural. This is due to the fact that metal is not hylomorphic, meaning that it is not passive enough to simply adopt any pre-existing form. It resists. Metals are hydraulic solids, and metallurgy can be considered a special kind of hydraulics that is employed for very slow and resistant movements. Moreover, metal is not primarily suited for the fabrication of tools, but rather finds its true purpose in the creation of weapons. It does not belong to the realm of State apparatus, but rather to the domain of the war machine. It introduces the concept of a metallurgical history that stands in contrast to the royal or State history. Only within the framework of the former do man and nature align on the same side. The genuine artisan, the nomadic one, is not the individual who shapes homogeneous matter according to a predetermined form, but rather the one who embraces the inherent flow of matter, forging it based on its distinct inclinations and singular tendencies. He takes part in the nature of invention, "nature as invention".[5] The metallurgist is akin to a stoic who lives and works in harmony with the creative and resilient forces of nature.

Intriguingly, in reference to the metaphysical concept of "panmetalism", Deleuze and Guattari propose that metal possesses a form of livingness, containing intensities that stimulate thought. While in *Difference and Repetition* thinking is portrayed not as the recognition of pre-existing objects, but rather as the active intervention in the formation of sense-data, in *A Thousand Plateaus* metallurgy emerges as the genuine science dealing with matter in its entirety:

> … thought is born more from metal than from stone: metallurgy is minor science in person, 'vague' science or the phenomenology of matter. The prodigious idea of *Nonorganic Life* … was the invention, the intuition of

[5] See the Appendix I on Max Ernst.

> metallurgy. Metal is neither a thing nor an organism, but a *body* without organs. (ATP, 454, italics in original)

Deleuze and Guattari discern a fundamental productivity of matter within metallurgy, operating at a level deeper than that of vegetable or animal productivity. "Because metal is the pure productivity of matter, those who follow metal are producers of objects par excellence" (454). Right from the outset of *A Thousand Plateaus*, Deleuze and Guattari allude to this productivity by referencing the molecular dimension of matter (though the structure of metals is more accurately described as sub-molecular). However, what distinguishes metallurgy in this context is the manner in which the molecular dimension becomes evident in conjunction with the history of technicity, and its interplay with States and nomads.

The primacy of metallurgy underscores the inherent connection between the capitalist machine and the war machine. While *Anti-Oedipus* defines machines in terms of production, *A Thousand Plateaus* portrays wars as a dual process of production and destruction. All machines are war machines because machines inherently stand in contrast to structures, and thus, are opposed to State apparatuses. How does capitalism assimilate wars as its inherent mode of production? To address this question, it's necessary to examine the form of lines of flight that define the capitalist machine in connection with modern manifestations of State and power. Eric Alliez and Maurizio Lazzarato discuss this fundamental question in their work *Wars and Capital* (2018). In this work, they put forth two central theses: firstly, they argue that the economy constitutes a continuation of war through different means. Secondly, stemming from this proposition, they suggest that every form of production encompasses an element of destruction. They write, under capitalism, "The machine of production is no longer distinguishable from the war machine integrating civilian and military, peace and war, in the single process of a continuum of isomorphic power in all its forms of valuation" (Alliez and Lazzarato 2016, 20). The intricate interconnection between economy and war gives rise to novel formulations of both concepts: economy encompasses not only production and profit but also destruction and deficit. And war is not defined only through armies but also by "the circuit of production, taxation, technological innovation, science, and employment" (43).[6] This aligns with the fundamental role of metallurgy in Deleuze and Guattari's perspective. Through metallurgy, war and economy, natural history and technicity merge into a unified entity.

The capitalist machine functions as a war machine, promoting the essence of freedom as unbounded power. Hence, in line with Alliez and Lazzarato's perspective, unbridled production becomes indistinguishable from unrestrained warfare, exemplified by the two World Wars and perpetuated through the Cold War, wars on terror, and more. Wars are anything but marginal to the logic of capital. The production of goods is merely an expansion of the production of bads ("*Warfare* largely anticipated and prepared *welfare*. Starting in the 1930s, the two became indistinguishable"; p. 21). Hence, alongside the workers of the world, there is also an invitation to the soldiers of the world, who should orchestrate a global coup d'état against the capitalist war machine. However, Alliez and Lazzarato are not entirely faithful to Deleuze and Guattari's domain of micropolitics, as they often oscillate, in the course of their work, between molar wars and molecular wars.

Alliez and Lazzarato's theory of pervasive global wars as the function of Capital's war machines, originally derived from Deleuze and Guattari's notion of the war machine, sheds light and elaborates on the idea of micro-fascism introduced in the ninth Plateau of *A Thousand Plateaus*. Alliez and Lazzarato write, "While the war machine of Capital governs through an 'inclusive' differentiation of property and wealth, the new fascist war machines function through exclusion based on racial, sexual, and national identity. The two logics seem incompatible. In reality, they inevitably converge" (28). We protest against racism and sexism, but no one seriously questions the subordination of the poor to the rich or the classification of society based on looks, talent, background, or any factor that grants privilege to one social group over another. Many of us consider these privileges normal or even natural. Micro-fascism is at work not only when we ignore the voices of those who are voiceless and

[6] This emerging battleground is evident within the coalition of influential nations in the global South known as BRICS, wherein Russia's prominence diminishes due to its adherence to conventional notions of warfare. This shift elevates China to a leadership position, as it perceives the realm of economics as the authentic theater of conflict with the west.

destined to remain silent and oppressed, but also when these people internalize their situation, accepting it as their identity: "It's too easy to be antifascist on the molar level, and not even see the fascist inside you, the fascist you yourself sustain and nourish and cherish with molecules both personal and collective" (ATP, 237). Even now, in every corner of the planet, we can easily find Antifa persons and collectives composed of fascist molecules.[7]

The mere existence of molecular fascism alters the conventional perception of totalitarian States and, to an extent, the operations of capitalism. In the concluding pages of the ninth Plateau, Deleuze and Guattari illustrate fascism through the lens of total war, thus setting it apart from totalitarianism. The distinction lies in the differentiation between the molecular war machine and the molar mechanism of power. "When fascism builds itself a totalitarian State, it is not in the sense of a State army taking power, but of a war machine taking over the State" (254). The war machine, far from being a centralized apparatus, is composed of lines of flight or quantum flows. Under fascism, the line of flight, the movement of deterritorialization, is inverted and becomes "a line of pure destruction and abolition" (254). This renders it antagonistic to life, imbuing it with a self-destructive, suicidal, nature: "so-called total war seems less a State undertaking than an undertaking of a war machine that appropriates the State and channels into it a flow of absolute war whose only possible outcome is the suicide of the State itself" (255).

What are the fascist elements of capitalism, their destructive molecules? Addressing this question requires us to turn to the concluding pages of the twelfth Plateau: "The factors that make State war total war are closely connected to capitalism: it has to do with the investment of constant capital in equipment, industry, and the war economy, and the investment of variable capital in the population in its physical and mental aspects (both as warmakers and as victim of war). Total war is not only a war of annihilation but arises when annihilation takes as its 'center' not

[7] Mainstream media and mainstream academia (particularly, academic philosophy) often use to label this stance as a kind of relativism, asserting that it blurs the categorical demarcation between moral and political good and evil. It might be relativistic in a sense, but it doesn't align with a conservative perspective toward the present state of matters. It is non-relativist in maintaining a clear differentiation at the molecular level between ethical and micro-political good and evil.

only the enemy army, or the enemy State, but the entire population and its economy" (ATP, 464). This stands as one of the primary sources that prompted Alliez and Lazzarato to formulate their concept of economic wars. However, their focus doesn't rest on the inherent similarity between fascism and capitalism. Although they affirm that "the 'nature' of fascism is not foreign to that of capitalism", but they focus on "the right to kill" and Foucault's account of biopower, instead of discussing the molecular destruction of life in both systems (Alliez and Lazzarato 2016, 202). In Deleuze and Guattari's analysis, what we deem most significant is the observation that capitalism globalizes the element of fascism at the molecular level, manifesting within the population and its economic fabric: to desire one's own repression; to willingly endure exploitation solely for the sake of participating in the consumptive game. Moreover, the consummation itself is a form of exploitation. It is through the total exploitation in the capitalist form of life that war and economy merge into each other. This signifies that the exploitation becomes absolute and war seeps into the daily existence of the populace through the conduit of the economy—the sorcery of the dollar. The result is not very different from the destiny of individuals under the fascist regimes: not solely those who meet their demise, but also the multitude who transform into mere applauders of the ruler's rhetoric, into vacuous images, into zombies. Capitalism brings about death in a similar fashion. Jim Jarmusch's film *The Dead Don't Die* (2019) effectively satirizes how capitalism transforms everyone into zombies, underscoring its destructive power that unveils its fascist nature. The underlying message is straightforward: we are already dead.

References

Alliez, Éric and Lazzarato, Maurizio. *Wars and Capital.* Translated by Ames Hodges. South Pasadena: Semiotext(e), 2016.

Bachelard, Gaston. *The Formation of the Scientific Mind: A Contribution to a Psychoanalysis of Objective Knowledge.* Trans. Mary McAllester Jones. Manchester: Clinamen Press, 2002.

Deleuze, Gilles and Guattari, Félix. *Anti-Oedipus: Capitalism and Schizophrenia*. Trans. Robert Hurley, Mark Seem and Helen R. Lane. Minneapolis: University of Minnesota Press, 1983.

Deleuze, Gilles and Guattari, Félix. *A Thousand Plateaus: Capitalism and Schizophrenia*. Trans. Brian Massumi. London and New York: Continuum, 2004.

Deleuze, Gilles. *The Logic of Sense*. Trans. Mark Lester. London: The Athlone Press, 1990.

Deleuze, Gilles. *Difference and Repetition*. Trans. Paul Patton. London and New York: Continuum, 2004.

Guattari, Félix. *Molecular Revolution*. Tr. Rosemary Sheed. Penguin, 1984.

Heidegger, Martin. *The Question Concerning Technology and Other Essays*. Translated by William Lovitt. Garland Publishing, 1977.

6

From Bio-Machines to Bio-Politics

In the film *Square* (Östlund, 2017), a curator navigates between two distinct realms: the polished professional world of art communities, gallery openings, and intellectual gatherings, and the brute and gritty reality of social conflicts, problems, and disorders. This film is an inquiry about the meaning of art by drawing a clear distinction between the royal or bourgeois art, subservient to the demands of State, and the molecular minor art that functions as a guerrilla force, or war machine, resisting these demands. The film's protagonist finds himself torn between the inescapable tumult of social-political issues that infiltrate every facet of private and public life, and a conservative intellectual community that vehemently opposes these challenges, resorting to feeble counter-strategies in a bid to preserve its existence. They prefer to avert their gaze from the harsh realities of social life and are willing to spend millions of euros to shield themselves from confronting the rudity and brutality of it. Through the eyes of the protagonist, we can think: "What strikes me is the fact that in our society, art has become something which is related only to objects and not to individuals, or to life. That art is something which is specialized or which is done by experts who are artists. But couldn't everyone's life become a work of art? Why should the lamp or the house be an art object, but not our life?" (Foucault 1984, 350). How can everyone's life

M. Parsa, *Machinic Ontology*, https://doi.org/10.1007/978-3-031-99561-3_6

become a work of art? Through a revolution, perhaps. But what kind of revolution? What is a revolution?

This conceptual distinction between two types of artists can be readily extended to philosophers, among others. It allows us to differentiate between conservative philosophers aligned with the State and those philosophers who function as a dynamic force, operating as a war machine, a dynamite, against it: those who maintain a separation between philosophical problems and the problems of life and those who facilitate their interaction, connection, or even fusion. There are only two types of philosophy: the royal philosophy of the State and the minor philosophy of life. In a limited context, this differentiation also reflects the distinction between analytical and continental philosophers. While there is a growing trend, especially among academic philosophers, to deconstruct the division between analytical and continental philosophy, we believe that, in a certain sense, particularly in the context of contemporary academic analytical philosophy, this distinction still holds significance. The first group of philosophers are workers (of the body of science, of the State), whereas the second group embodies both creators and destroyers (the true philosophers of free will). The first group extracts philosophical problems from the lifeless pages of textbooks, the rigid structure of science, and established philosophical traditions. In contrast, the second group derives their problems from lived experiences, societal dynamics, scientific experiments, and living traditions. For them, tradition is not just something to abide by but to embody, just as science is not merely studied but lived within.

This leads us to explore the connection between philosophy and life, and subsequently, to address the question of what relevance non-organic, molecular, or machinic life bears on ethical-political existence—how individuals and communities live—and how it relates to the overarching purpose and definition of philosophy. Is there an inherent link between pure life and philosophy? In the context of human existence, how can philosophy facilitate the convergence of our lives with life as the fundamental productive essence of being, essentially merging anthropology into ontology? The transition to ontology cannot occur without traversing the realms of brute philosophy and art, as well as the raw, unrefined life depicted in works like *Themroc* (1973) and the philosophy/life of the

Cynics. As these examples illustrate, any shift towards ontology necessitates the shift from morality to ethics, from the act of choosing among existing lives to the endeavor of inventing new forms of life.

What kind of life are we living? We select from existing forms of life to determine our path. This choice serves as both the limitation and the definition of our freedom. We are free to choose between ready-made lifestyles. This represents the ethical counterpart to the ontology of substances, which defines reality in terms of pre-existing objects. However, inventing how to live always appears as a peculiar thing, transgressing or violating the boundaries set by the State's definition of human freedom. To live differently, one must discard all existing forms of life, challenging the conventional essence of being human. This entails merging with the ontological core of life, the untamed and pristine life of nature. Creation inevitably involves destruction. Introducing the concept of non-substantial life, the non-organic life discussed in previous chapters, must resonate ethically and politically in the endeavor to invent novel forms of social existence. To achieve this, philosophy must engage with the concept of pure life at an ethical level—a life devoid of existing structures, a purely material life, without organs and organizations. As will be explored further, this is indeed the case with the Cynics, as investigated by Michel Foucault. The central theme of this investigation asserts that the philosophical inquiry into the meaning of life is far from being merely theoretical; it is intricately related to the essence of political power, which imposes pre-existing forms of life for consumption.

The Self-Maintenance of the Social Machine

Foucault's later philosophy, especially in his final seminars at the Collège de France published as *The Courage of Truth*, can be understood as exploring a connection—or rather an identification—between life and truth. This connection transforms the nature of both terms. The only authentic life is the true life, the animal life exemplified in the Cynic form of life, and the only truth that exists, distinct from scientific or democratic truth, is the truth of this life. Foucault begins his lectures in *The Courage of Truth* with the concept of parrhesia, understood as the practice of truth-telling about oneself. This leads to a conception of truth that encompasses the non-apophantic logos. However, starting with the ninth seminar (February 29, 1984), he turns to the Cynics, who integrate truth-telling into the form of life (*Courage*, 173). With the Cynics, the non-apophantic logos becomes a form of life. Parrhesia involves the courage to tell the truth in the face of the prince or tyrant, but since doing so is always dangerous (58), the reference to the Cynics illustrates that the only way to be truly parrhesiastic is to live a distinct life: the Cynic must improvise their life at every moment. They tuned the music of life in a jazz mode.

What is the cynic way of life? The Cynic is often seen as antisocial. Diogenes, the most famous Cynic, is renowned for his arrogance and egoism. What Foucault's reading reveals, however, is that the Cynic is antisocial without possessing an ego, because the dog has no ego. Recalling our chapter on Simondon, one could argue that the Cynic embodies the pre-individual sociality of pure life. The Cynic is not merely a dog; rather, he transforms animality into a gesture—indeed, a political gesture. It materializes "existence as an *oeuvre*", as a work of art, as something that unfolds on a stage. The barrel of Diogenes is, in fact, the stage of a theater. And, it is a profoundly political theater. The Cynic's antisociality is the resistance of pre-individual sociality against individualist sociality. Diogenes' indifference in his confrontation with Alexander the Great symbolizes political activism in a situation where open and direct politics is impossible. Foucault's ultimate aim in referring to the ancient Cynics is to suggest that the emergence of reluctant forms of life in the twentieth century, ranging from hippies to Western Buddhism, reflects such a political gesture. It is about how to live in your solitude as if you are on the stage of a theater—the theater of existence—making your life your philosophy and your way of living your teaching. Or perhaps the other way around, bringing your solitude onto the stage of the theater: "the philosopher eats with great gluttony, he stuffs himself; he masturbates in public, regretting that hunger cannot be so easily relieved; he does not condemn incest with mother, the sister, or the daughter; he tolerates cannibalism and anthropophagy – but, in fact, he is also supremely sober and chaste" (LoS, 130).

In *The Courage of Truth*, Foucault examines the Greek concept of "self-care" (epimeleia) as an imperative and seeks to make sense of it. The crucial question is: what is the "self" in "take care of your 'self'"? In his reading of Plato's *Laches*, Foucault defines the "self" in terms of the connection between truth and life, or, in terms of true life (*Courage*, 160–161). One must take care of oneself because one's life is at risk within the false life of society. The transition from Socratic to Cynic epimeleia leads to a peculiar notion of the self, one that is identical with the dissolution of the self or selflessness. Cynic epimeleia means we must take care of our animality. Diogenes is the true political animal.

The political significance of Cynicism leads Foucault to link the nakedness, shamelessness, poverty, and animality of Cynicism to what he calls "the militant life" (283). He even suggests the political dimension of animality by noting that the Cynic dog can be "the guard dog which accosts enemies and bites them" (284). Every dog is a guard dog, just as every machine is a war machine. Or, in the spirit of Deleuze and Guattari's *A Thousand Plateaus*, every dog is a pack of wolves, because we are not dealing with individuals, but rather, as Jacques Rancière emphasizes, with the communities of ordinary people—the restless mass, the mob. The goal of politics is to transform the ordinary people from herds into packs, as Deleuze and Guattari write, "In becoming-wolf, the important thing is the position of the mass, and above all the position of the subject itself in relation to the pack or wolf-multiplicity" (ATP, 32). This is how, according to Foucault's reading, Cynic individualism takes on a profound political significance.

It is life that must take care of itself through political communities. The guard dogs, or packs of wolves, assume this responsibility. This is an inverted reading of Hobbes, transforming the meaning of war into an act that turns toward nature, rather than arising from it. Man is a wolf, not per se, but rather in order to care for some "self". The relationship of self to self is only possible through the militant life—rebellion against existing forms of life and transforming life itself into an aesthetics of existence. The social machine cannot function without this epimeleia, without the constant social war that sometimes manifests in the form of politics. This is why Foucault insists that resistance is the origin of power (Foucault 1997, 167). Resistance constitutes the ties, bolts, and nuts of the social machine. The "self" of the social machine is nothing other than life. This is the aesthetics of existence, insofar as creativity is inherent to life. As Deleuze asserts in his seminar "What is a Creative Act?", held in Paris on March 17, 1987, the creative act is defined through the act of resistance: "there is a fundamental affinity between a work of art and an act of resistance" (Deleuze 2006, 322–323). When the mode of life itself becomes a work of art, resistance assumes the form of an aesthetics of existence.

This is, indeed, the ultimate meaning Foucault assigns to immanent ethics. The core of ethics is psycho-political resistance against one's own desire—a resistance of a desire against the desire shaped and manipulated by the social structure. As I discussed elsewhere (Parsa 2024), the immanence of ethics rests on the unity of life and truth. For Nietzsche, the evaluation of values suggests that the moral criterion (truth) is inseparable from the evolution (life) of the moral system itself (264). Nietzsche's critique of formalism (see Nietzsche 1974, 205–206), particularly his critique of Kant's formal morality, leads him to incorporate living drives, or libido, into the moral system, thereby distancing ethics from Kantian conformity to social norms. For Nietzsche, real critique—what he calls "philosophizing with a hammer"—involves breaking (with) social norms and their truths, the truths of consensus. As Deleuze notes, this is what "remains at the heart of the Kantianism like the worm in the fruit: true knowledge, true morality, and True Religion" (Deleuze 1983, 90). For Nietzsche, life is the true *a priori*, the truth itself, "what makes anything become different from itself" (Parsa, 265). This leads to the identity of ethics and ontology, making the "self" identical with the thing-in-itself. Through this reading, the Foucauldian *epimeleia* becomes the core of ontology and logic—the relationship between the thing and its "self"—making possible the identity of the "self" of *epimeleia* and the selflessness of the Cynics. It is no longer about the moral self-care of the subject, but rather the ontological self-maintenance of the social machine. This is, in fact, an affirmative critique of Kant that leads to a Kantian ontology: incorporating the second Critique into the first in order to conclude with the third Critique, *The Aesthetics of Existence*, culminating in a non-apophantic ontology.

The Crisis of Democracy

Foucault frames the problem of democracy in terms of the issue of governance. The problem, according to Foucault, is that all forms of governance, including the democratic one, imply obedience. This also applies to self-governance, or more fundamentally, to self-command. In this sense, the concept of freedom in democracy, defined as the governance of people over themselves, becomes a point of reflection. Parrhesia challenges the very structure of command and obedience, and its ethical core involves a deconstruction of self-command and self-obedience. On the other hand, democracy is based on consensus, while parrhesia, particularly Cynic parrhesia, disrupts, challenges, and unsettles this consensus. This is why Foucault insists that parrhesia is dangerous for the democratic city (*Courage*, 36–37), and democracy is dangerous for parrhesia (45). The question now is: is a politics without commandment and obedience possible? A politics that does not generate obedient subjects and communities? How can we approach the political realm, if not through self-commanding individuals and communities? As Giorgio Agamben suggests and Catherine Malabou elaborates in her recent book, a politics liberated from self-commandment and self-obedience would have an anarchic nature.

The Athenian democracy, which Foucault regards as the true democracy (*Courage*, 42), is characterized as a regime where decisions are made not by an elite few, but by the majority. The majority serves only their own interests. "They want to command (*arkhein*) by themselves" (42), meaning they seek to serve their own interests. The crisis of democracy should be understood in terms of what Foucault refers to as "the crisis of Parrhesia". Parrhesia demands courageous individuals and communities—those who question the rulers without fear. This could be mistakenly interpreted as having the courage to demand one's own interests from the rulers, the courage to assert one's own rights: slaves fighting for the rights of slaves, blacks for the rights of blacks, women for the rights of women, and so on. However, Foucault, from the outset, connects parrhesia to the very form of life, the aesthetics of existence, imbuing it with an ethical meaning that distinguishes it from the interests of any particular group: slaves striving not to remain slaves, blacks not to remain blacks, women not to remain women, what implies a radical change in the very structure of social life. This is what leads Foucault to connect two meanings of parrhesia: as the truth of speech and the truth of life. In simple terms, within democracies, there is freedom of speech for obedient subjects and communities to express their interests and rights, but not freedom of life, the freedom to invent one's own form of life. The reason is that the former implies a subject or community that is governable, while the latter implies an ungovernable subject or community. In this sense, we can trace the essence of Foucault's critique of democracy through Canguilhem's reexamination of life as monstrous and pathological, in contrast to the normal. Here, the pathological aligns with the ungovernable, and bios is placed at the center of politics.

Let us consider this simple formula: Democracy is good for the interests of individuals, but is not good for the disinterested life. Or, in terms of the identity of life and truth with which we began our reading of Foucault, it is not good for truth: "Either democracy or truth-telling" (*Courage*, 45). What does this formula mean? The issue lies in the relation of the subject to the pack and a philo-political notion of freedom. Foucault notes that democracy is not only about comfort and laziness, but also about obedience, subjectivation, and flattery. It prevents the formation of courageous, truthful, and free individuals. Parrhesia defines not only the characteristic of truth but also that of freedom, as it involves both telling the truth and speaking freely (39), or, in the case of the Cynic, living freely.

What alternative exists to the democratic structure of commandment and obedience? In her chapter on Foucault's anarchism, Catherine Malabou explores the idea that the later Foucault emerges as a proponent of anarchism. To do this, she follows Deleuze's approach by first reading Foucault through the lens of immanence, before expressing her dissatisfaction with this interpretation. For Deleuze, the core of the doctrine of immanence is the non-existence of the "inside". There is only one singular realm, and that is the outside. How can we understand Foucault's notion of the "self" in light of this Deleuzian conception of immanence? What does the "self" mean in the pure outside? According to Malabou, the Deleuzian interpretation of Foucault's concept of the "self" can be understood through the idea of "auto-affection" (Malabou 2023, 130). This leads to a form of anarchism, because "a fundamental form of auto-affection, government of self, construed as a fold, loses all hierarchical dimensions" (132). First and foremost, what Deleuze learns from Spinoza is that, in the outside, everything is an affect. A key characteristic of affect is that it cannot form consensus. This is evident in Deleuze's reading of Kant's third *Critique*, where he prioritizes the sublime as the foundation of Kant's critical philosophy: all harmonies are rooted in a radical disconsensus or tension. Malabou's reading takes a different direction. She equates immanence with auto-affection, and, according to her, "the logic of commanding and obeying" remains intact in the reading of Foucault through the lens of immanence. She argues that the doctrine of immanence implies "a domination of oneself". In contrast to this quasi-Deleuzian interpretation, Malabou argues that there is a deeper Foucauldian anarchism that implies absolute non-governability.

We are not convinced that Deleuze's concept of immanence implies domination, auto-affection, or the logic of commanding and obeying. Malabou's discussion of this issue is brief, focusing on Deleuze's readings of Spinoza and Plotinus. She does not address Deleuze's depiction of immanence as the pure outside, *a* life, in *Immanence: A Life* (Deleuze 2001), nor his interpretation of the Kantian Sublime in *Kant's Critical Philosophy* (1967). We find it odd that Malabou interprets Deleuze's ideas of affection and the fold as aligning with Plato's concept of the divine self-reflection of the soul in *Alcibiades*, when she writes, "In an economy not so far removed from the one Deleuze described as affection of essence, the soul reflects on itself by illuminating its divine part, as if folding itself in order to better capture the light" (Malabou 2023, 136). Furthermore, we find her claim ungrounded that "any affect theory, whatever it is, is *necessarily* rooted in auto-affection" (141). In our view, both Foucault (in his reading of Plato's *Laches*) and Deleuze (particularly in *Immanence: A Life*) define the self through a radical interpretation of life, or what can be understood as pure, unindividuated life. This indicates a moment in Deleuze's thought that has nothing to do with auto-affection, but only to affection pure and simple, a radical form of selflessness. The culmination of Foucault's thought in his later seminars, as illustrated in his reading of the Cynics, is the concept of epimeleia without self—a form of care that is free from subject-object relations. The Cynic takes care of his non-existent self, a gesture that is consistently accompanied by mocking and undermining the robust self of his opponent, Alexander the Great.

Whether in agreement with or against Deleuze, Malabou interprets Foucault's concept of revolutionary or militant life in terms of anarchism (140–141). Foucault's anarchism is inherently intertwined with his critique of democracy in *The Courage of Truth*. The so-called inner selflessness rejects any external hierarchy, including the democratic one. Democracy, which normalizes and makes transparent the structures of domination and obedience, is subverted by the Cynic's devaluation. Every democracy is built on micro-hierarchies both within individuals and between them. Challenging these micro-structures demands a radical position, which Foucault identifies in Cynicism.

What are these democratic micro-fascisms within and between us? In *Late Fascism*, Alberto Toscano argues that fascism and democracy (which, for him, always means liberal democracy) form a unified nexus through the concept of race:

> Of note and relevance for contemporary concerns is the specific light that the prism of race – of racial domination and racial capitalism –sheds on the nexus of fascism and democracy, and how it can help us to interrogate and displace the normative conviction regarding the absolute antithesis between fascist despotism and liberal democracy. (Toscano 2023, 42)

For Toscano, there is a "white supremist democracy", and he refers to "the democratic survival of fascism" or "The persistence of fascism within democracy" (89). According to him, fascism can be understood as a manipulation of individual freedom, or what Deleuze and Guattari articulate as the politics of desire. Fascism often takes the following form: *Now that I happen to be white, why not be proud of it and take advantage of it? Now that I belong to the superior race, why should I feel ashamed for insisting that my race is superior?* This fascist tendency arises from a mechanistic manipulation of life, suppressing its creative, monstrous core and transforming it into mere survival, conservation, and a selfishness that masquerades as freedom. In this way, individuals' desires are manipulated so that they come to perceive their servitude and obedience as freedom.

How can one resist this tendency? How can we confront the micro-fascisms within us? What kind of anarchism is needed to free ourselves from democratic domination and obedience?

How to Make Yourself an Ungovernable Dog?

Terrence Malick's film *A Hidden Life* (2019) offers a picturesque contribution to the Foucauldian distinction between true and false lives. It tells the story of Franz Jägerstätter's revolt, a defiance that leads him to embrace a concept of freedom in opposition to the "self". At the outset, Franz is far from being a militant. He is a content family man with a simple life, while the world around him is dominated by militarism. The people of the village must adapt to the overwhelming new reality. They are compelled to speak of and believe in race and blood. The external reality must be internalized within the minds and desires of individuals. Franz resists. He resists the social consensus, against a Demos that happens to be fascist. Religious belief gives him the strength to relativize the values of the sovereign structure and its economy. But the pressure is intense. He loves his wife, Fani, and his life. The pressure comes from two sides: not only the torture in prison and the brutal military structure, but also the words of the villagers, Fani's sister, his mother, and even the doubts and thoughts of Fani herself. These two forces are always intertwined; they must work together. Fighting against the latter is more difficult. Against the former, Frank finds himself free in prison. Against the latter, he finds himself free only when his wife says, "I'm with you always. Do what is right".

The true life must remain hidden. Franz is far from a hero. He is a martyr for nothing. People who want to help him remind him that none of the authorities are aware of him. His resistance has no impact on the course of the war. They don't understand what he is fighting for. Why, as quoted from George Eliot at the end of the film, is the good of the world dependent on the inner, hidden life of a forgotten villager? What is the good of the world? For Franz, it is not a question. He strives to take care of his "self", his parrhesiastic truth. All he knows is that lying to himself is the worst thing he could do. He is not trying to save the world from evil; if you want to save the world, start with yourself! In a desperate attempt to save him, the priest says, "say the oath and think what you like". Franz knows it is not that simple. His life is not separate from the situation he is in or the events that unfold around him. He must suffer and die; he must sacrifice. And this must happen through a trial, similar to Socrates', confronted with Athenian democracy. Franz already knows his fate. He recognizes the defeat without being defeated; it is better to be a victim of injustice than to perpetrate it. And the world is unjust. There is a struggle out there.

The economy of symbolic life (bios) can only be destroyed through sacrifice, by what Giorgio Agamben refers to as *homo sacer*. The painter who depicts the comfortable Christ, with a halo above his head, cannot convey the suffering of the real Christ or the life he endured. The painter says, "someday I might have the courage to venture, not yet. Someday I'll paint the true Christ". Painting the true Christ is only possible by living his life, something that requires courage. Franz makes his life a bare, killable life. This is his sacrifice, his way of becoming sacred. In sacrifice, the form of life and religious belief or philosophical thought become identical. If philosophy is the art of learning how to die, then philosophy as a way of life would be sacrifice.

People in the village say that what Franz does is out of arrogance. Fani's sister says it is pride. This is how it is seen from the perspective of economy. But Franz has no ego. The one who sacrifices himself cannot have an ego. As the priest says, his sacrifice benefits no one. This is why people mistake it for arrogance—because one sacrifices himself for the benefit of others. Franz's sacrifice is for the sake of bare life. It is neither for his ego nor for other individuals, but for the pre-individual core of ethics and politics. The political lesson of this sacrifice is that the consensus between egos, the core of democracy, is built on hypocrisy and role-playing to attain a symbolic position in society—selling your "self", your soul, to gain social status and reinforce the ego.

Bare life is animal life, what Agamben refers to as Zoë. The only way for Franz to remain human is by becoming a killable dog. He clearly reveals the political nature of cynicism. Sacrifice and resistance, the dog and the wolf, become identical: fighting against injustice without becoming unjust. Or, in other words, sacrifice is the transformation of Zoë into bios. This means that the Cynic's a-sociality and indifference reveal another level of sociality—the animal or bacterial sociality—that should be emphasized if one wants to consider alternatives to the current individualized sociality.

Not only bacterial sociality, but also viral, machinic, or cosmic. Franz is not just a killable dog, the Zoë, but also the mountain. The film begins and ends with the mountain. This reminds us that Franz's struggle is not about becoming human, but rather a struggle for existence. It is the mountain that exists. The mountain is where being and action, potentiality and actuality, coincide. Bare life finds its true meaning when it transcends the biological and acquires an ontological significance. Fani's certainty gives Franz strength: "I meet you there in the mountains". But when Franz is gone, Fani is left with a question: "Are you sure?" This song by David Lynch captures the essence of Terrence Malick's *A Hidden Life* perfectly:

> "Stand up and be a mountain
> Power of the dream
> The sea, it being a living thing
> River, like a hand
> Jump up, and be a cloud again
> Raining upon the man
> Oh, are you sure?
> Oh, are you sure?
> …"

Biopower and Bare Life

The governing power that once disciplined individuals as legal subjects now merges with life itself, embedding itself within the forms of collective life. A legal mechanism is now intertwined with a bio-machine. Foucault refers to this as the reflection of biological history on political history. The outcome is biopolitics, which no longer concerns individuals but populations, treating them as a living entity—the social body. This demands a particular form of power whose application shapes the social structure of life.

For Foucault, biopolitics is defined as the administration of life, with bio-power being the form of power that facilitates this administration. It is a power that governs not only the minds and subjects but also the social body, the population, and its modes of living. It is a defining feature of late capitalism, where politics becomes a form of economics—the management or administration of the house. For Foucault, life is not a natural force that remains untouched by, or external to, the techniques of power and knowledge (what Agamben refers to as Zoë, or bare life). Instead, life is directly manipulable through these economic techniques (bios). In this regard, life is nothing more than forms of life. At this point, technique and nature, technique and life, merge with one another. As Foucault explains in "The Right of Death and Power over Life", the population has two poles: it can be seen either as a machine or as a living body:

> One of these poles – the first to be formed, it seems – centered on the body as a machine: its disciplining, the optimization of its capabilities, the extortion of its forces, the parallel increase of its usefulness and its docility, its integration into systems of efficient and economic controls… The second, formed somewhat later, focused on the species body, the body imbued with the mechanics of life and serving as the basis of the biological processes: propagation, births and mortality, the level of health, life expectancy and longevity, with all the conditions that can cause these to vary. (*The Foucault Reader*, 261–262)

Specifically, Foucault examines the function of bio-power under neoliberalism, or late capitalism, as a means for "the controlled insertion of bodies into the machinery of production and the adjustment of the phenomena of population to economic processes" (263). The question is: what form of life does a population live, and why? What determines it? Bio-power transforms bodies into useful instruments, whose life, and the quality of that life, must be protected and enhanced. This represents a new form of labor exploitation that regulates individual freedom in service of the machinery of production. We can no longer discuss natural freedom in terms of contingency, controlled by power; instead, it is a complex metabolism of regulated freedom. The feeling of freedom and security in individuals functions as a driving force for economic growth and is therefore confined within the limits of its governmentality.

Foucault compares these two forms of power by declaring that the disciplinary power regulates the individuals by the menace of death; if someone transgresses the laws of the sovereign power, he would have to face his death. But bio-power aims to regulate life continuously instead of leaving it free when it does not transgress the sovereign power and its laws. In Foucault's term, under the operation of bio-power, the law becomes the norm and institutions become apparatuses (266). The term "apparatus" is a translation of Foucault's "*dispositif*", which posits him in the tradition of the objectification of politics that begins with Auguste Comte and entails replacing the government of things for the government of people (see Thomas Lemke, pp. 88–90). For Foucault, this does not mean making people the governable things, but alters the sense of governmentality and makes it a contribution in the disposition of *things*, instead of applying a sovereign power from without. But this changes also the nature of the political *objects*. Humans gain their objective agency, their machinic collectivity, in becoming the *dispositive* agents.[1] This underlies a machinalization of social life. This is similar to the way Deleuze, in "Postscript on the Societies of Control", distinguishes between disciplinary societies and control societies, where in the former the sovereign power applies only in the case of threats and transgressions while in the latter the points of control are scattered everywhere and regulate the flux of life. This indicates a difference between two types of social machines; in the first type the natural vitality and the mechanistic structure are separated while in the second type they are merged into each other. In Giorgio Agamben's term, under the capitalist religion, their "profanation" is almost impossible and this leads to a shift from the classical process of subjectification into a desubjectification, which is, melting all subjectivities into the overwhelming network of apparatuses.

Agamben's critique of Foucault in *Homo Sacer: Sovereign Power and Bare Life* rests on the argument that neoliberal regimes, which Foucault views as administrators of life (bios), also engage with bare or killable life

[1] Thomas Lemke in *The Government of Things*, makes a similar attempt to reconciliate the subject and the object of political governmentality, but his focus on the *thingness* prevents him to reach the full collective agency of non-subjective operators. Thus, by adding a nuance to Thomas Lemke's view, we claim that Foucault's notion of governmentality does not concern the humans-things assemblage, but rather humans-machines (Lemke 2021, 84–85).

(Zoë). According to Agamben, Foucault failed to recognize the continued existence of traditional forms of sovereign power within modern states:

> Foucault never brought his insights to bear on what could well have appeared to be the exemplary place of modern biopolitics: the politics of the great totalitarian states of the twentieth century. The inquiry that began with a reconstruction of the *grand enfermement* in hospitals and prisons did not end with an analysis of the concentration camp. (Agamben 1998, 119)

Modern regimes operate by incorporating traditional methods of sovereignty as exceptions. However, it is this necessary exception that defines the characteristic nature of these regimes. Agamben himself refers to camps (concentration camps, refugee camps) as examples of this state of exception. It is a structure that reveals itself through the consideration of the Nazi regime as an exception in the twentieth century. This structure can also be seen in the way political regimes in countries like Iran, while emphasizing the ideological nature of their power structures, support their propaganda—seen in the educational system or mass media—with military intervention. This includes the deployment of armed moral police, the execution of media activists, the shooting of peaceful protesters, and so on. This structure is also evident in Franz Fanon's emphasis that colonialists, whose goal appears to be the peaceful spread of the Western way of life in "uncivilized" parts of the world, inevitably resort to brutal military force to achieve or secure this objective. These examples demonstrate that the exception alters the nature of the original goal or intention. The necessary combination of soft power with hard power signifies an inherent structure that operates within all neoliberal regimes, even if sometimes in a concealed form.

Agamben's main goal in *Homo Sacer* is to clarify the convergence of two mechanisms of power that operate together in the modern state: traditional sovereign (juridico-institutional) power and modern biopower (Agamben 1998, 5). For Agamben, this convergence is only possible through the inclusion of bare or killable life into the political realm, or the "politicization" of bare life (8). We can interpret this inclusion as an ontological move, since the incorporation of bare life into the political realm also represents an extension of the political sphere to encompass anything external to the social-political machine.

The question now is: Is the animal Cynic life, as described in *The Courage of Truth*, Bios or Zoë? We believe that Agamben's critique of Foucault's biopolitics does not apply to Foucault's concept of Cynic parrhesia. If homo sacer is excluded from the political realm and is killable, the Cynic presents a political gesture—a political mannerism—that challenges this structure. First and foremost, the Cynic is non-economic, non-administrable, and therefore non-governable. Agamben's "inclusion" of the exception into the social machine, or its "extension" to the ontological level, implies a radical transformation of the dispositif, or apparatus.

The dispositive subjectification of humans is always unfinished; therefore, political states must resort to combining it with traditional totalitarian techniques as necessary exceptions. However, no complete synthesis or fusion is possible. The juridico-institutional power and the dynamics of life remain somewhat separate, though deeply entangled. This entanglement gives rise to a mechanism in which regulatory apparatuses replace abstract and direct laws, aiming to control all flows and movements. Even in these societies, two meanings of freedom can be easily distinguished: on one hand, individual freedom, which is targeted by juridical power and its laws, and on the other hand, a flexible mechanism that regulates the population while maximizing respect for individual freedom within the confines of its rules. This flexibility is the core meaning of neoliberal freedom. It differs from natural contingency, although it appears strikingly similar to it. Here, we encounter a kind of correspondence between the concepts of machine and apparatus (dispositif) when considering the social machine. An apparatus is the result of the synthesis between the disciplinary machine and the control-based machine, combining the mechanistic social machine with the pure bio-machine. The notion of apparatus suggests the fragmentation of the social machine into smaller pieces that can be carried by each individual in their pockets. Instead of large institutions designed to regulate individuals (like Hitchcock's battles between man and man-made structures), we now have apparatuses in our pockets that control us: social media has taken the place of the medium of society. This is why, as Michael Hardt and Antonio Negri argue in *Empire* (2000) and *Multitude* (2004), capitalism requires no central unity or unified essence and can instead be understood in terms of multiplicities. In a Hegelian move, we begin with the unity of nature and its continuous flux, move through the dualism of the sovereign institution of disciplinary power and subjectified individuals, and ultimately arrive at a new unity—this time in the form of univocity—in which the dualism of controlling power and subjectified individuals multiplies throughout the system.

Giorgio Agamben, in *What is an Apparatus?*, separates the apparatuses and the living beings (or living substances) (Agamben 2009, 13). But at the same time, he defines "the extreme phase of capitalist development in which we live as a massive accumulation and proliferation of apparatuses … we could say that today there is not even a single instant in which the life of individuals is not modeled, contaminated, or controlled by some apparatus" (15). Here, we can clearly see the structure in which the social machine is not totally merged into a natural living being but is multiplied into small couples of living individuals and their apparatuses, a situation which is similar to Deleuze's scattered control points that govern the entire flow while leaving *everywhere* the spaces of freedom. With Agamben, we find first a separation of the apparatus (*Gerät, appareil*) and the human user, and then their unification in the way the human society operates as a dispositor, which is to say, an economical-political system of power. This implies a transition from the limited sense of machine, namely the machines that we use, toward the ontological sense of it, namely the society as a machine and at the ontological limit, the whole nature as a machine, although neither Foucault nor Agamben do not emphasize on this ontological level. They focus rather on how this social-ontological sense returns back to the relationship between the individuals and the apparatuses that they use, while deconstructing their instrumentality. In this way, two meanings of the term technology, the "technologies of the self", and the machinic meaning of technology join together. The objective is to provide a description of the society which is not structural, as Foucault insists on the strategic sense of the dispositive opposed to the essence of a structure.

As Agamben mentions, apparatus has at once three senses, a juridical sense, as the enactment of the social law, a technological sense, as the arrangement of a machine or mechanism, and a military sense, as the operationality of a war machine (Agamben 2009, 7). Thus, there are three ways that the apparatuses which are coupled with individuals can connect them together, as obeying subjects, as producing or desiring machines, and as war machines. Agamben insists that these three meanings are indeed one and the same meaning that articulates itself in different domains. In other words, it is the same machine that operates within the juridical, the technological, and the military domain. The machine of governance that produces subjectifications is also a system of desiring machines that, although are controllable by their beloved devices, but are capable to transform the hybridity of these devices and their small desiring machines into a war machine through a strategic guerrilla logic that connects them (the hybrid man-apparatuses) together and forms a monster-mega-machine. In simple terms, where there is power, there should also be resistance. And the complexity of sovereign power merged into bio-power renders the essence of resistance complex and hybrid. This is radically different from any one-directional sovereignty and any theoretical effort to limit the meaning of militarism to the state. The same holds to the meaning of government, whose essence must be searched in the ethical self-government. The incompleteness of the synthesis provides then an essential heterogeneity which is indeed the point of radical and primordial resistance. This heterogeneity is the element of power therefore resistance and power are never separable. This entanglement marks the Foucauldian account of power and the structuralist ones. It marks also the hybrid nature of natural machines as well as social machines. They are not the conscious subjects, or any revival of subjectivities, that provide the resistance against the sovereign power, but rather the monstrous and heterogenous society of machine-man hybrids, the bio-apparatuses, the Cyborgs, or the community of computer viruses, as Deleuze thinks, or collective hack programs, that go out of control through their collectivities.

This is not politics in the conventional sense, but an ethico-political life which is revolutionary in its forms of living. If politics is a continuation of war with other means, with Foucault the war, which is disseminated in the molecular level, is transformed into micro-technicity. The concept of dispositive in the sense of strategy gains now the new meaning of using bio-technologies to politicize the molecular wars. Hence, society is not a peaceful organism, but a war machine although the war can be transformed strategically in many forms of molecular politics. This modifies drastically the Hobbesian view of society by changing and expanding the meaning of wars and conflicts with the help of the strategic sense of the dispositive. In this sense, there is no form of life which is not political. If war is the essence of life, as Heraclitus and Nietzsche recognized, and life is inherently social, then politics constitutes the forms of life. Not only politics integrates the human technicity in the essence of the political animal and in this way dehumanizes it, but also is extended to biopolitics, and at the limit, the political ontology.

References

Agamben, Giorgio. *What is an Apparatus? And other Essays*. Translated by David Kishik and Stefan Pedatella. Standford University Press, 2009.

Agamben, Giorgio. *Homo Sacer: Sovereign Power and Bare Life*. Trans. Daniel Heller-Roazen. Standford: Standford University Press, 1998.

Deleuze, Gilles. *Pure Immanence: Essays on A Life*. Trans. Anne Boyman. New York: Zone Books, 2001.

Deleuze, Gilles. *Nietzsche and Philosophy*. Trans. Hugh Tomlinson. London: Athlon Press, 1983.

Deleuze, Gilles. (2006). "What is the Creative Act?" in *Two Regimes of Madness*. Ed. David Lapoujade. Translated by Ames Hodges and Mike Taormina. Semiotext(e) Foreign Agents, pp. 312–324.

Deleuze, Gilles and Guattari, Félix. *A Thousand Plateaus: Capitalism and Schizophrenia*. Trans. Brian Massumi. London and New York: Continuum, 2004.

Deleuze, Gilles. *The Logic of Sense*. Trans. Mark Lester. London: The Athlone Press, 1990.

Foucault, Michel. *The Courage of the Truth (The Government of Self and Others II) 1983–1984*. Trans. Graham Burchell. London: Palgrave Macmillan, 2011.

Foucault, Michel. *Foucault Reader*. Edited by Paul Rabinow. Pantheon Books, New York, 1984.

Foucault, Michel. 1997. "Sex, power, and the politics of identity," in *Ethics: Subjectivity and Truth*, pp. 163–73.

Lemke, Thomas. *The Government of Things: Foucault and the New Materialisms.* New York University Press, 2021.

Malabou, Catherine. *Stop Thief! Anarchism and Philosophy.* Translated by Carolyn Shread. Polity, 2023.

Nietzsche, Friedrich. *Gay Science.* Trans. Walter Kaufmann. New York: Vintage Books, 1974.

Parsa, M. (2024). "Immanent Ethics and Deconstruction". *Angelaki, 29*(1–2), 263–274. https://doi.org/10.1080/0969725X.2024.2322298.

Toscano, Alberto. *Late Fascism: Race, Capitalism and the Politics of Crisis.* London and New York: Verso, 2023.

7

Conclusion: Dream-Machines

Two modes of thought can be distinguished, a poetical thought that has its eyes to the future, and a conservative, immunist or mechanistic thought which builds its road on the ground of past sediments. The defining characteristic of the first mode is its monstrosity. A future that is not monstrous is not a future, but merely a reflection of the past. These two modes of thought can be exemplified in terms of two functions of the brain, namely dreaming and remembering; that which liberates and that which binds; that which deterritorializes and that which reterritorializes. The social machine must be considered like a collective brain that does not operate only as a collective memory but rather as a collective dreaming-machine. Dreaming is the highest degree of thinking and every machine is a dream-machine. Only under this condition can an artificial brain truly function as a brain. It is only in this condition that history opens a place for revolution, as Marx writes, "The social revolution of the nineteenth century cannot draw its poetry from the past, but only from the future" (Marx 1997, 329). This distinction also highlights the difference between two types of abstraction: conceptual or categorical abstraction and abstraction as deterritorialization. It's time to recognize that we can be dream-machines. All war machines are dream-machines. We can plug everything in the world into ourselves in order to produce a

M. Parsa, *Machinic Ontology*, https://doi.org/10.1007/978-3-031-99561-3_7

schizophrenic dreamer like Daniel Paul Schreber. A day-dreamer in the manner of David Lynch. Everything can be an apparatus in our pocket and we are able to transform each apparatus into a weapon. This is the true meaning of digital revolution; it is not a technical or scientific revolution, but a true political revolution. The true political subject is the dreaming collectivity which is composed of micro bio-machines. This is how a people can truly dream and realize its dream into *a* life. Machinic ontology underlies the ideal of an autonomous society which operates without any representative governmentality, without any gap between the governing class and the self-governing collective singularities. If capitalism takes humans as mechanical machines, or mere instruments, machinic ontology implies self-operating natural societies whose governance issues from the collective articulation of natural flows of desire. A self-governing collectivity that announces the full identity of ethics and politics.

Capitalism colonizes our dreams. As Mark Fisher in his *Capitalist Realism* declares, capitalism announces the realism of "there is no alternative". We became realist, in the sense that we turned our dreaming machines off. The result is the dead societies composed of working-class democratic individuals who consume and wheelwork. It is in such societies that authoritarian measures are "implemented within a political structure that remains, notionally, democratic" (Fisher 2009, 1). As Fisher, referring to *Children of Men* (2006), elaborates, it is possible, and even plausible, that from a democratic system remains only the protectionist sectors, namely the military and police functions. In this context, communism heralds rather, in our understanding of the term, the possibility of collective dreaming as an alternative basis for the social life. Idealism indicates the power of dreaming. Even the so-called German Idealism cannot exist without this collective aspect. But it's a while that we forgot to dream collectively. We live in an era of realism, a capitalist realism. What took the place of collective dreaming is the individual wish for success, the competition or battle for success, which is playing the game of capitalist instrumentalization. Success underlies here nothing but being a good and useful instrument for a machinery that generates infinite types of class conflicts. In the current state of affairs, there is no way to say that something is beautiful without taking some other things as ugly, no way to get success without dooming some others as looser. We hate each other even when we play the game of love. We hate even our (pseudo-)heroes at the same time of respecting them. Our heroes are indeed "paid wage-labourers".[1] Yes, we all sold our souls to the devil. What is missing among us is the man who sold the world in order to gain back, not his individual soul, but the spirit of the world.

Our previous chapter concluded by exploring the entanglement of two modes of power or social structuration. This entanglement reflects the extension of labor into free time in the digital era, marking a shift from the traditional Marxian concept of the machine—limited to factory work—to a new conception of the machine that encompasses the entirety of social life. Neoliberalism is a regime in which exploitation is woven into the very fabric of individual freedom. Our conclusion is grounded in the idea that every machine can be transformed into a war machine. Therefore, the late capitalist society, as a digital machine, can also be understood in terms of its potential for resistance and revolution. This represents the ultimate meaning of collective dreaming: digital dreaming rather than digital fantasizing. This slogan has two dimensions: First, there is an inherent connection between unconscious desire and collective dreaming. The machinic ontology suggests that the collective or the socius dreams prior to the individual. In this sense, Werner Herzog can film the dreams of the green ants, the Amazonian jungle, or even the galaxies. Second, daydreaming differs from fantasizing: the former involves thought, while the latter suggests a readiness for manipulation.

Capitalism underlies a globalized awareness. Communism is demanding that such awareness should entail a collective dreaming. The machinic ontology implies that there is no limit for this demand. It is the unconscious that dreams and the unconscious is never individual. The machinic unconscious marks the transcendental ability of flows to join and to be articulated in contingent manner. The whole story is about opening up the possibility of alternatives. Now, after the advancement of capitalism

[1] "The bourgeoisie has stripped of its halo every occupation hitherto honoured and looked up to with reverent awe. It has converted the physician, the lawyer, the priest, the poet, the man of science, into its paid wage-labourers" (Marx and Engels 1970, 34).

into a unique digital network, the alternative takes the shape of a global digital unconscious.

This is compatible with two forces that are distinguished in the *Communist Manifesto*. On the one hand, bourgeoisie (the ruling class under capitalism), in the "need of a constantly expanding market for its products", chases itself "over the whole surface of the globe" (Marx and Engels 1970, 35). This indicates the globalist nature of capitalism that disrespects the national and cultural borders. There is an important anti-nationalist force in capitalism: "In place of the old wants, satisfied by the productions of the country, we find new wants, requiring for their satisfaction the products of distant lands and climes. In place of the old local and national seclusion and self-sufficiency, we have intercourse in every direction, universal inter-dependence of nations. And as in material, so also in intellectual production" (35). On the other hand, the bourgeoisie, by subjecting the country and rural areas to the rule of the towns and cities, "has agglomerated population, centralised means of production, and has concentrated property in a few hands. The necessary consequence of this was political centralisation. Independent, or but loosely connected provinces, with separate interests, laws, governments and systems of taxation, became lumped together into one nation, with one government, one code of laws, one national class-interest, one frontier and one customs-tariff" (36). In this way, it strengthens the national borders and nationalist efforts to centralize the political power. Capitalism has an international side and a nationalist side. Marx and Engels gear a battle in two fronts against capitalism. On the one hand, they intend to build an emancipating force on the globalist aspect of bourgeoisie. Where exploitation is international, there would be a need for international revolutionary resistance against it. On the other hand, they insist on local, but non-nationalist, demands and experiences which are no less important in constituting the emancipating power. And even Marx and Engels (long before Derrida) are well aware that this is not a battle with two fronts, but more about the way the bourgeoisie deconstructs itself from within.[2]

[2] "The productive forces at the disposal of society no longer tend to further the development of the conditions of bourgeois property; on the contrary, they have become too powerful for these conditions, by which they are fettered, and so soon as they overcome these fetters, they bring disorder into the whole of bourgeois society, endanger the existence of bourgeois property. The conditions

Because it is the proletariat, the product of the bourgeoisie, that burdens the emancipatory force. Similar to the capital itself, the proletariat resists against its abstraction and formalization, against becoming "an appendage of the machine" (39). The global capitalism is a machine that condemns the workers (namely, everyone) to be mere instruments. The communist alternative heralds a new idea of machine, by introducing a new conception of labor. Along with the emancipatory globalism in the *Manifesto*, there is an emancipatory machinism which is fueled by the molecular war machines whose elements are far from instrumental. It is now the multitude as war machine: "Masses of labourers, crowded into the factory, are organised like soldiers" (40). A machine works, and labor, like money, can be abstract. It is no longer wage labor, but the social labor of life itself. The capitalist abstraction is the classical, substantialist abstraction that unifies completely the work of the worker and the amount of money that is equivalent to it. But abstract machines can become war machines, in the sense that, the movement of abstraction emerges from the multiplication of the singular though collective forces in the pre-individual state of social life. In this sense, the abstract things are not equal. Abstraction is not a movement of identification, but conversely a movement of deviation and singularization. This is the meaning of the new alliance that Marx and Engels through the slogan of "the workers of the world, unite!" cry. This is a cry to build a new alternative war machine, a new machinic unconscious. This is an alliance for the sake of freedom and singularity: "You have nothing to lose but your chains!" How an alliance can donate freedom to the participants? How a machinism can generate something other than exploitation? How a commonality can be compatible with singularity? It is only possible if we can liberate the concept of machine, and together with it, the notion of work and co-working, from mechanism and structuralism which are indeed new versions of unity and identity.

Guattari and Negri in *Communists Like Us: New Spaces of Liberty, New Lines of Alliance* introduce the thought of "a whole new way of working together" (Guattari and Negri 1990, 13). On this basis, they "define

---

of bourgeois society are too narrow to comprise the wealth created by them" (Marx and Engels 1970, 38).

communism as the collective struggle for the liberation of work" (Guattari and Negri 1990, 10). Their essay elaborates implicitly, under the title of communism, a synthesis between the ideas of the machinic unconscious and the war machine. There is a way to work together, in a machinery, that leads to liberation instead of exploitation: "A work which no longer is crystallized in the form of private property, which does not consider the instruments of production as ends in themselves, but as means for attaining the happiness of singularity and its expansion in machinic rhizomes - abstract and/or concrete" (40). In a manner which is familiar to us, Guattari and Negri oppose two concepts of machine to each other. On the one hand, capitalism is a machine because its essence is work; it is a machinery for subjugating individuals to the exploiting work. On the other hand, if communism provides an alternative, it should reformulate the essence of work, and this is exactly what Guattari and Negri, in *Communists Like Us*, pursue to do with recourse to "machinic rhizomes". It is a book on machinic communism. Guattari and Negri declare that communism should introduce a new concept of machine. The machine that liberates versus the machine that exploits; the emancipating labor versus the wage-labor; the rhizomatic machine versus the mechanistic machine. In the capitalist system, labor is mere wage-labor, which is to say, it is regulated by the abstract value that is countable, payable and compensable with money. Now, the question, or rather the task, is to replace abstraction in the sense of deterritorialization as an instrument to evaluate labor on the singular basis, so that the co-working does not underlie a reduction to the abstract wage-labor, but rather the singular value that emerges from the actual nature of the work, which is creativity. In this context, Guattari and Negri redefine communism: "communism must be more than just the sharing of wealth (who wants all this shit?) – it must inaugurate a whole new way of working together" (13). A working together which is not organized by the mechanistic power-structure of capitalism, a co-creativity which is rooted in the creative pre-individual sociality of nature, its essential rhizomatic machinality. In the current situation in which economy and politics are merged more than always at the global level, it is the structure of power, the relation between ethics and politics, and most fundamentality, the nature of onto-logic that must be changed. What is needed is a genuine social agency, a true social process of decision-making. If the logic of multiplicities entails reasoning without universals, a multitude can function without being reduced into a herd, a *Volk*, a *demos*. The onto-logic of multiplicities, which is free of universals and substances, demands a social life without identities of any kind, no matter national, European or sexual, a new economy which is not based on the false identity of capital, and a politics which is not brutally representational. All of these separate the structure of power in social life from the machinic metabolism of life. An appeal to a non-representational democracy lays on the collective nature of the life itself:

> Communism is nothing other than a call to life: to break the encirclement of the capitalist and socialist organization of work, which today leads not only to a continuing surplus of repression and exploitation, but to the extinction of the world and humanity with it. (Guattari and Negri 1990, 11)

Work should be rooted in the functionality of life, or in our term, in the machinality of nature, particularly in the current situation in which work and life are no longer separate. The capitalist machine shapes our lives even when we are not at work, exerting its disciplinary and biopower through mass media. These media are no longer confined to the times we sit in front of a TV, use a laptop, or check our smartphones; they now accompany us in all moments and activities. We live online. Time, for us, has become nothing more than waiting for the next vibrating notification. We are nothing but ad-viewers. Our freedom is not from or against the power-structure but an essential feature of the functioning of the structure. What we earn in this market is the individual freedom and the immunity or security that protects our imaginary egos. Because desperate egos are good workers. They work more when they are more desperate, sometimes at traditional work places, sometimes as ad-viewers. Capitalism colonized our dreams, after subjectifying our beliefs and loves into "the semiotics of capital". We are lonely individuals who hate each other while playing the game of love that we learned from cheap products of capitalism. Guattari and Negri announce their desire for a new mode of social life which is not structured by the capitalist or socialist power. It does not oppose the individual to the collective: "communism is the establishment of a communal life style in which individuality is recognized and truly liberated, not merely opposed to the collective" (Guattari and Negri 1990, 16). If communism is a communal working that does not exploit the workers and respects their individuality and creativeness, its globalism is far from the capitalist application of a form of life to different cultures. It underlies at once singularity and collectivity. It sets individual freedom against collective, yet singular, freedom. Freedom to choose and freedom to create and dream collectively. Freedom to consume commonly and freedom to generate collectively.

We shouldn't mistake the sustainability and persistence of the natural machine for the invincibility of the capitalist machine. Rigid power-structures are the most fragile entities. Capitalism, with the help of its apparatuses and media, confuses two notions of the machine and steals the power and flexibility of the natural machine in order to generate the illusion of "there is no alternative", our capitalist realism.

Last but not least, the machinic ontology does not lead to the dissolution of the political subject into a mere hallucinatory pre-individuality which is doomed to the natural machine while losing any kind of agency. It is rather about displacing the agency from the imaginary ego to the collective-emotive share of individuals. It is about basing politics on real desires, the natural machinic desires, instead of structural needs and wills which are merely imposed by and are the mirror image of the sovereign power ("*ich will*"). It is the share of the creative and productive collectivity in each individual that harbors the real political agency. It can also be called subjectivity, if one understands subjectivity in terms of singularity instead of fixed individuals:

> Communism has nothing to do with the collectivist barbarism that has come into existence. Communism is the most intense experience of subjectivity, the maximization of the processes of singularization – individuation which represent the capability potential of our collective stock. No universality of man can be extracted from the naked abstraction of social value. (39)

References

Fisher, Mark. *Capitalist Realism: Is There No Alternative?* Zero Books. 2009.
Guattari, Félix and Negri, Toni. *Communists Like Us: New Spaces of Liberty, New Lines of Alliance.* Translated by Michael Ryan. Semiotext(e), 1990.
Marx, Karl. "The Eighteenth Srumaire of Louis Bonaparte", in D. McLellan (ed) *Karl Marx: Selected Writings,* Oxford: Oxford University Press, 1997.
Marx, Karl and Engels, Fredrick. *Manifesto of the Communist Party.* Peking: Foreign Languages Press, 1970.

Appendix A: Max Ernst and The Schizoanalysis of Nature

Max Ernst is the artist who investigates the intricate connection between the autonomous or machinic aspect of human mind, known as unconscious, and the creativity of nature. While Dadaism focuses on the notion of automaticity or machinism, Surrealism on psychoanalytic operations of unconscious, and natural sciences on the marvel of nature itself, Ernst's art orchestrates an intrinsic communication between all these realms. The main aim of this appendix is to conclude from this that he is the genuine artist of the exteriority of the unconscious desire. In two volumes of *Capitalism and Schizophrenia*, Deleuze and Guattari examine the notions of machinic unconscious, desiring machines, desiring production and war machines. In the following, we endeavor to trace these concepts throughout Ernst's body of work and illustrate how his artistic endeavors aptly embody the themes pursued by Deleuze and Guattari in their collaboration.

In Ernst's work, the unconscious is rendered petrified and stratified, yet this petrified unconscious leads to a de-objectification, clearing a space for the idea of natural deep history. In this way, Ernst makes everything a fossil. However, it is a living fossil that unravels itself throughout

M. Parsa, *Machinic Ontology*, https://doi.org/10.1007/978-3-031-99561-3

natural history. He utilizes the artistic technics in such a way that the consciousness of the artist becomes passive to the inventiveness of the unconscious. In Ernst's work, from one painting to another, from sculpture to sculpture, the unconscious dons different masks repeatedly, never revealing its face. This involves a transformation of imagination, a transformation that is, in fact, an externalization. In *Beyond Painting*, Ernst famously states, "artistic expression is the visualization of the imagination, the transformation of the inner world into the outer world" (Ernst 1948, pp. 8–9). Elsewhere, in a 1961 interview with the BBC, he elaborates on this idea, extending it beyond a one-directional transformation:

Seeing usually means that you open your eyes to the outside world. It is possible to see another way; you close your eyes and you look into your 'inner world'. If you can make a synthesis of these two important worlds you come to result in what can be considered as the synthesis of objective and subjective life.[1]

The inside doesn't solely pertain to an individual's mind; it can also encompass culture and technology, the intersubjective inside, whereas the outside denotes nature itself. This synthesis constitutes the center of Ernst's art. His work endeavors to embrace the wild, incorporating it into the inner world and capturing the dreams and nightmares of the external nature. It can be described as a psychoanalysis of nature. But the term psychoanalysis, through Freud and Lacan, remains limited to the intersubjective realm of language, falls short to deal with the external nature and deep history. This is exactly what brings us to Deleuze and Guattari's critique of psychoanalysis and their invention of schizoanalysis. In their view, the former is structural while the latter is machinic, which is to say, at the level of the wild, the unconscious produces its objects.

Ralph Ubl, in *Prehistoric Future: Max Ernst and Return of Painting between the Wars*, discusses the notion of "unconscious production" in Ernst's work through references to Sigmund Freud and Walter Benjamin, with a passing reference also to Jacques Lacan (Ubl, 6–11). He distinguishes unconscious production from the concept of "artwork as artifact", which essentially involves conscious creation evident in graphic art, cubism, and the technique of collage. After studying Benjamin, Ubl

[1] https://elephant.art/5-questions-laurence-owen/.

concentrates on how Max Ernst's work embodies a resurgence of painting and its spontaneity, contrasting with the perceived decline of painting in cubism and graphic art. Hence, Ernst, who was initially intrigued by the idea of the "death of painting" before the First World War, rediscovers the spontaneous vitality of painting after the war. Here, one can discern two perspectives on technicity or automaticity, each aligning with distinct views of the unconscious: Firstly, there is what corresponds to the empty and lifeless reproduction of forms, and secondly, what corresponds to the inventory and innate productivity of matter. In essence, we are dealing with two types of machines: artifact machines and natural machines. According to Ubl, Ernst's work, particularly his rediscovery of painting, can be summarized as a dedication to natural machines or the autonomy of nature. The experience of war led Ernst to draw parallels between wild nature, human beings, and artifacts, viewing them all as monstrous machines.

The dichotomy between artificial and natural automaticity proposed by Ubl prompts us to reinterpret Ernst's psychoanalytic art through the lens of Deleuze and Guattari's concept of the machinality of nature. Ernst was indeed influenced by Freud, yet his fascination with nature propels his art beyond the confines of Freudian and Lacanian psychoanalysis. Before delving into Ernst's art, let us briefly outline the principal features of Deleuze and Guattari's conception of the unconscious, which forms the foundation of our interpretation of Ernst's work.

Firstly, the unconscious is machinic, characterized by a fundamental automatism. This concept, as presented in *Anti-Oedipus*, serves as a rebuttal to Freud's renowned analogy of the unconscious as a theater stage. Deleuze and Guattari contend that the unconscious is not like a theater that represents its objects; it is rather like a factory that produces its objects.[2] However, it's crucial to pay close attention to the meaning of production, considering the Benjaminian critique of artistic production and the dichotomy between artifact machines and natural machines that

---

[2] This stance also opposes Lacan's structural interpretation of the unconscious. Deleuze and Guattari instead offer a dynamic and pragmatic understanding of the unconscious, one that is constantly engaged in action. Ian Buchanan, in his commentary on *Anti-Oedipus*, notes this structural inclination in Freud himself, particularly evident in his early works such as *Interpretation of Dreams* (see Buchanan, 32).

has been highlighted. Deleuze and Guattari's unconscious factory doesn't churn out uniform and useful objects; rather, it operates as a natural factory, producing without adherence to quality control: "a classical theater was substituted for the unconscious as a factory; representation was substituted for the units of production of the unconscious; and an unconscious that was capable of nothing but expressing itself—in myth, tragedy, dreams—has substituted for the productive unconscious" (AO, 24). This contrasts with Freud's theatrical model of the psyche, which corresponds with his focus on myth and tragedy, particularly Sophocles' *Oedipus Rex*, to understand unconscious desire (See Holland 1999, 21).

Secondly, the unconscious is external; it doesn't pertain to the unconscious mind but rather to nature itself.[3] The connection with nature lies in how schizophrenia becomes the principal manifestation of the operation of the unconscious. According to Deleuze and Guattari, schizophrenia, conceived as a model of the unconscious, extends beyond mere mental disorder to signify nature as a process of production. In their perspective, nature exhibits a schizophrenic quality, characterized by its wild and fantastical production: "What the schizophrenic experiences, both as an individual and as a member of the human species, is not at all any one specific aspect of nature, but nature as a process of production" (AO, 3).

Thirdly, the unconscious involves a particular sense of abstraction intertwined with the concept of the war machine, the political assertion that comprehending the unconscious requires a revolutionary action. This concept is primarily developed in the second volume of *Capitalism and Schizophrenia*, namely *A Thousand Plateaus*, where they contrast the war machine with the State apparatus (ATP, 388–389). While "the concern of the State is to conserve" (394), a war machine is essentially revolutionary because the war machine is "a pure form of exteriority, whereas the State apparatus constitutes the form of interiority" (390). Given the inherently non-peaceful nature of these movements, abstract machines can be regarded as war machines, functioning in accordance with the

[3] Ray Brassier in his chapter "Concrete Rules and Abstract Machines: Form and Function in *A Thousand Plateaus*" in *A Thousand Plateaus and Philosophy* recognizes this aspect by making a link between "the auto-construction of the real" and "the machinic unconscious" (Brassier, 267).

principles of "guerrilla logic" (19). This signifies a war not only against social domination, but also against the psychoanalytic domestication of desire—a rebellion not just for revolution, but also for revolt.[4]

Nature as a factory for generating non-utilizable entities, which stands in opposition to the economy of the State apparatus. Schizophrenia as a residue of nature within human bodies, inciting rebellion against societal norms—a veritable war machine. These are the Deleuzo-Guattarian concepts that guide us in interpreting Max Ernst, the artist who bridged the Dadaist-Surrealist rebellion with the creative forces of nature. For Deleuze and Guattari, art does not exist to imitate nature (ATP, 5). It engages in natural production, a participation that is anything but peaceful.

Nature as Invention[5]

Art is not a human endeavor; it is inherent to nature itself. Max Ernst's art strives to present nature as the fundamental source of all artistic activity. This applies to art in its diverse forms, encompassing industry and machinery. In his art, Ernst attempts to dissolve the figure of the artist into nature, presenting nature itself as the primordial artist. In Deleuze and Guattari's term, Ernst's art displays the "fantastic factory of Nature and Production" (AO, 49). In doing so, he establishes a connection between artistic creativity and natural creation, a revolutionary act in art history so that a German newspaper even accused him to "plagiarism" from nature (Warlick, 42). As André Breton asserts in reference to Ernst, the artist is not the master of his work, but functions as a medium (Breton, 61). This implies an identification between nature and the human psyche, claiming that our mind's artistic creativity is a continuum of the marvelous creativity of nature itself. The artist invents along with

[4] Commentators often seek Ernst's rebellious attitude in his relationship with his father. For instance, Ian Turpin writes, "Looming large in his memory was the figure of his father who, both as a deeply religious man responsible for Ernst's strict Catholic upbringing, and as a Sunday painter, provided a model against which he was to rebel" (Turpin 1979, 3).

[5] This title refers to "Max Ernst und Die Natur als Erfindung" [Max Ernst and Nature as Invention], an exhibition of Ernst's works in Kunst Museum Bonn from 12 October 2022 to January 2023, which centers his *Histoire naturelle* for understanding the totality of his work.

nature. This is why Ernst is fascinated by the technique of frottage or decalcomania. In this manner, he generates textures regardless of how the artist perceives them. Only afterward can the artist intervene and behold the product of his art. He withdraws to reveal the creative force of nature as the true source of multiplicity. Similarly, Ernst observes that when the artist recognizes analogous forms across various aspects of nature, it's because a singular force invents both the dead and the living, the sky and the earth, the microscopic and the macroscopic. It's not merely about recognizing forms within a decalcomaniac chaos, but rather the actual emergence of forms from within that chaotic milieu. Ernst pioneered the non-analogical theory of art, echoing the thesis of the univocity of being in the history of philosophy. Imagination is the actual expression of nature. The contrast between Max Ernst and his Bonn friend, the expressionist painter, August Macke, can be likened to the distinction between Duns Scotus and Thomas Aquinas. In this sense, Ernst is the artist depicting the univocity of nature.[6] The idea of univocity is apparent in the following statement of him, when describing a catalogue that he found in a print shop and then used in numerous collages:

> One rainy day in 1919, finding myself in a village on the Rhine, I was struck by the obsession which held under my gaze the pages of an illustrated catalogue showing objects designed for anthropologic, microscopic, psychological, mineralogical and paleontological demonstration. I found there joined together elements of figuration so remote that the sheer absurdity of this assemblage provoked an acute intensification of my visionary faculties …. (Ernst 1948, 14)

This not only entails the exploration of collage as an anti-painting, anti-modernist trend but also carries a naturalistic philosophical connotation. The fusion of collage and Dadaist automaticity in Ernst's oeuvre yields a conception of the unconscious that resonates with Deleuze and Guattari's *Capitalism and Schizophrenia*: machinism rather than organicism, embracing a non-organic universalism. This assertion is applicable not solely to Ernst's machines like "Little Machine Constructed by Minimax Dadamax in Person" (1919–1920) but extends to the entirety

---

[6] Turpin explains the contrast between Ernst and Macke in terms of Dada's attack on western culture, avant-guard versus bourgeoisie (Turpin, 4).

of his artistic output and his approach to art. On this basis, we argue that, on the one hand, Ernst's work is philosophical in nature, as it offers a universal perspective not only on artistic innovation but also on human invention, perception, psychoanalysis, history, and so forth. On the other hand, Ernst's work embodies the experimental approach of a natural scientist, providing the inherent machinism of his art. As Turpin highlights in a comparison with the prominent Cubist painter, "Where Picasso claimed to find rather than to seek, Ernst's oeuvre is characterized by an attitude of enquiry", functioning "as a means of investigation" (Turpin, 3). In this manner, he embraces both philosophical unity and scientific diversity, a working-together of organs devoid of central organicity.

Jardin Peuple de Chimeres (1936) holds significant importance in the context of this notion of univocity. We recognize human or animal faces within botanical life. For Ernst, the crucial point is not that we tend to perceive familiar forms everywhere, in clouds and stars, and occasionally in flowers and gardens. This isn't about our imagination or our subjective perspective; what's at stake here is nature's productive imagination. There are no analogies in nature, only multiplicities. Everywhere in nature, the presence of circles and spirals, the resemblance between suns and eyes, follows a simple principle: the Earth and the sun share a specific relationship. Gravity, encompassing atomic and subatomic electrical forces, as well as weak and strong forces, stands as the fundamental force of nature, shaping the emergence of all forms, both living and non-living. This shift represents a departure from the conventional anthropological theory of art, which centers on the creativity of the artist, and instead embraces an ontological theory of art that focuses on the creative natural force responsible for generating forms and information. This aligns with the fact that Ernst is not only interested in nature and its diversity, but also in technics and machines. This includes fantastical and useless machines, as seen in *Little Machine Constructed by Minimax Dadamax in Person* (1920), as well as explorations of animal-machine hybrids, as exemplified in the famous work *Elephant Celebes* (1921). To Ernst, art, as the genuine work of nature, also encompasses technique, and his *Natural History* (1926) includes also the history of technology.

Ernst's work can be seen then as a comprehensive exploration of natural history in a general sense. In fact, in historical terminology, Ernst aspires to be the painter of deep history, which is what lies beyond our known history. He paints the planet, this very planet, and its evolution, devoid of any human observers. This is why his art can be regarded as a phenomenological inquiry into the things themselves. What persists in the absence of human observers, categories, and objects? How does nature invent its objects? Ernst starts with a basic line that divides the earth from the sky. This marks the inception of his art, aligning with the very beginning of nature's artistry. Naturally, this perspective is from Earth's vantage point. Nevertheless, it is at least an objective perspective.[7] Next comes the sun, the perfect circle, and then, on the earth, the chaotic wild from which dark objects strive to reach the bright perfection. This is Ernst's simple depiction of natural history. He aims to differentiate between how nature expresses itself and how human consciousness recognizes it, prioritizing the former over the latter. His paintings of this planet exist outside the confines of our measurable time scale. The works that belong to the *Natural History* collection (1926) particularly *La Mer et la pluie*, *Un coup d'oeil*, *Le Châle à fleurs givre*, and *Le Tremblement de terre*, indicate the deep natural history, the *Urgeschichte*. There are also a few of his works that indicate specific places but offer no clues about the time period, which holds significant importance. This is exemplified in works like *Jerusalem* (1962), which portrays a specific place but certainly not within the timeframe of known religious events. There's a touch of humor here in the transformation of such a sacred place for human history into a geological reality. Even the painting titled *The Twentieth Century* (1955) can be interpreted equivocally; it alludes not only to the destructive aspect of our twentieth century but also to the concept of the twentieth century not in human terms but rather on a deep or absolute scale. Therefore, one could regard Jerusalem not as Celsius zero (in terms of the calendar) but rather as Kelvin zero (physicogeological). And the piece from 1955 is the Kelvin twentieth century. The same applies to Ernst's flora and fauna. The emergence of what is commonly known as life, the organic life, isn't a revolution in the planet's evolutionary force. The eyes

[7] In contrast to Hegel and the longstanding tradition of Platonism in philosophy, Ernst perceives objectivity not through universality but rather by approaching the perspective of non-categorical objects. In this regard, Ernst's conception of objectivity aligns more closely with Nietzsche's perspectivism.

aren't defined solely by their function or instrumental operation, but rather by their formation, their solar function, and their inherent natural technicity (*Solar Currency-System, The Wheel of Light, The Figurative*). Science has often exaggerated the distinction between the living and non-living, but art has the ability to transcend scientific dualisms and embrace ontological monism. Even the paintings of humans in the *Natural History* collection are dehumanized; they are completely dissolved into the faceless interplay of lights and shadows (*Eve, The Only One Left to Us*). Humans are transmuted into geological forces. The prominent motif here is the sun. Whether it's an earthquake, the flora or fauna, birds in the wild, or a mother and child, everything occurs under the sun, on this planet. Long before the invention of cages, birds in the wild were trapped, living yet petrified (*The Origin of the Clock, The Conjugal Diamonds*). For Ernst, birds represent life itself, and their so-called petrification is merely the machinization of life, which is, its liberation from the living substance. It is not life that emerges out of soil, it is rather the primordial wild life that is petrified in its evolution, under the sun (*Petrified Forest - Forêt pétrifiée* 1927, *The Great Forest—La grande forêt* 1927, *Petrified Life - La vie pétrifiée* 1933). With Ernst, the dead nature is animated, while the living is petrified, under the sun. The sun and the earth, everything gains its form and matter from them and their synthesis.

Ernst's works do not represent nature, they don't leave it intact, but rather apply a natural violence upon it. They disrupt its serene existence, employing a raw, natural violence where the inner beast clashes with the outer, and vice versa. While Francis Bacon, in Gilles Deleuze's reading, depicts body's spasm, "*the action of invisible forces on the body*" (Deleuze 2003, 41), Ernst demonstrates well the spasm of nature, the action of invisible forces of nature on nature. This encapsulates the core essence and purpose of art, demonstrating why it is inherently intertwined with nature. The key point here is not that the bird captures a photo or image of the sun, but in the sun violently imprinting its mark upon the bird, leaving an indelible stigma (*L'oiseau soleil*, 1955). The sunflower is the stigmatized sun. In this context, Ernst's artwork, especially *Jardin Peuple de Chimeres*, serves as both a homage to and an extension of Henri Rousseau (1844–1910), most notably his masterpiece *Forêt vierge au soleil couchant* from 1910. While Rousseau maintains a separation between the sun, the wild, and the human within his synthesis, Ernst seamlessly amalgamates them into a colossal machine-nature. This signifies a shift from the tranquility of Rousseau towards Ernst's portrayal of violence and intensity. The result is epitomized in *Le roue du soleil* (1926), where the sun transforms into a machinic component, and life on Earth manifests in its most tumultuous and obscure form. This destructive force is indistinguishable from the generative force, as Ernst portrays both the ruins and the emergence of civilization as interchangeable entities. In *L'oeil du silence* (1943–44), it is impossible to determine whether they represent ancient origins of archaic civilizations which are emerging from natural chaos or the remnants of future cultures after devastating wars or natural catastrophes. Evidently, they epitomize the same essence, the same metabolic cycle.

Natural History and Deep History

Stratified Rocks, Nature's Gift of Gneiss Lava Iceland Moss from 1920 establishes a connection between natural history, as portrayed through the layers of stratified rocks, and a disorganized or fossilized horse body, linked to manmade instruments of measurement. This suggests that the horse's organs and surroundings are reorganized to create a new monstrous natural machine. Moreover, Ernst's use of the overpainting embodies the stratification in both nature and artification. As Ubl mentions, "*Stratified Rocks* is not a collage – it's an overpainting; and even when erroneously taken for a collage, it does not convey the impression of having been produced with freehand cuts. The rigidity of the space and the tension of the layers suggest not so much the agility of a knife or scissors, but instead directionless forces working against each other (sedimentation, glaciation, stratification)" (Ubl, 41–42).

Sedimentation, glaciation, and stratification not only characterize Ernst's painting technique but also underscore the intrinsic relationship he establishes between natural history and artification. Appearing Deleuzoguattarian, these are geological concepts employed to depict living beings and technical instruments. This leads us to consider the concept of deep history, which pertains to a historical era beyond the reach of our conscious awareness. This concept would assist in emancipating Ernst's idea of natural history from the confines of positivist natural sciences, imbuing it with the contingency and creativity inherent in artistic pursuits.

Daniel Smail and Dipesh Chakrabarti are two of the contemporary thinkers who have studied the implications of the notion of deep history. Contrary to the notion of prehistory, as utilized by Ubl for understanding Ernst's work, the notion of deep history has more unconscious denotations, implying a split between the deep past and the surface of the present. It refers to a past that affects the present only unconsciously. Smail focuses on the biological nature of this affection, while Chakrabarti emphasizes the geological aspect of deep history, two directions which are evident in Ernst's work. This difference in focus makes Chakrabarti to criticize Smail's *On Deep History and the Brain* for reducing the human geological force to a (neuro)biological framework.[8] In "The Brain of History, Or, the Mentality of the Anthropocene", Catherine Malabou attempted to reconcile these two accounts of deep history by introducing a concept of the brain of history that should not be confined solely to the human or animal brain (Malabou, 49–53). Malabou's reconciliation resonates with Ernst's method of crafting collages from biological and geological forms, as seen in works like *The Entire City* (*La ville entire*, 1935), and emphasizes the concept of the external unconscious. In this context, natural history can be likened to a giant brain that generates diverse forms. In this manner, it encompasses both the remote past (devoid of any human traces) and the distant future (after the cessation of humanity). This concept serves as a foundation for unifying various phases of Ernst's oeuvre, spanning from his natural history collection and fascination with biological and geological phenomena to the pervasive influence of industry and culture evident in his work. Ernst's art encompasses everything, whether it's the lifeless Earth, a bird in a natural cage, a human figure seen from behind, or a complex hybrid machine, all within the framework of natural deep history.

---

[8] "But it is the history of human biology and not any recent theses about the newly acquired geological agency of humans that concerns Smail" (Chakrabarti, 206).

In this respect, Ernst's work holds relevance in contemporary discussions surrounding the Anthropocene, notably through his exploration of the concept of war in a broader context—a destructive force that transcends mere human agency. Malabou's notion of the brain of history as a notion that encompasses not only the deep past and deep future, but the entirety of history, including the period of human existence, implies a radical modification of the notion of human agency. It prompts us to perceive human life and civilization as petrified and fossilized. It surpasses human consciousness or limited biological agency and encompasses humans as geological forces. This can be accomplished by embracing Chakrabarti's geological approach and incorporating Smail's perspective on history as a functioning brain. Ultimately, Malabou's primary objective was to provide a fresh perspective on history by bridging the longstanding gap that exists in our understanding of the relationship between history and nature. As Malabou asserts, "nature has a history, it is history". Chakrabarti's first thesis in his essay is straightforward and unambiguous. He boldly announces "the collapse of the age-old humanist distinction between natural history and human history" (Chakrabarti, 201). An important conclusion of this would be the extension of the destructive forces of nature to human agency, as exemplified in the notion of Anthropocene. This sheds a light on Ernst's depiction of war, in works such as *Europe after the Rain II* (1940–42), where the ruins of the war appear as geological structures. The petrified forest is indeed a ruined city. This can be precisely understood in terms of the collapse that Chakrabarti observes: the dissolution of the distinction between natural history and human history is what unites natural destructive forces (*Tremblement de Terre*, 1926) and human destructive forces, echoing the deleuzoguattarian notion that all machines are war machines.

Machine-Monsters and War Machines

Machines have consistently held a central place in surrealism, forming a defining characteristic of the movement as seen in the works of Marcel Duchamp and Man Ray. The surrealist depiction of machines leads to the introduction of a concept of machine without utility and beyond human objectives, a concept that is perfectly compatible with Deleuze and Guattari's view of machines in *Anti-Oedipus*, where they introduce nature as a huge industry that produces not for the sake of utilization:

"we make no distinction between man and nature: the human essence of nature and the natural essence of man become one within nature in the form of production or industry, just as they do within the life of man as a species. Industry is then no longer considered from the extrinsic point of view of utility, but rather from the point of view of its fundamental identity with nature as production of man and by man" (AO, 4).

Non-utilizable machines lack teleology and hence lack organicity, as nature, operating as a giant machine, produces without purpose. Ernst shares this view to machines in his works. He created plenty of non-utilizable and non-organic machines. Nevertheless, what makes him particular among his surrealist colleagues is that he, as a prominent Dadaist, introduces the concept of war into this dynamic, proclaiming that all machines are inherently war machines, as Ernst in an interview in 1958 says "Dada was a bomb".[9] In doing so, he imparts a profound political significance to the artistic movement. Similar to Deleuze and Guattari's perspective in *A Thousand Plateaus*, where the concept of a war machine implies invention and nomadic movements capable of dismantling the State apparatus and its laws, or being co-opted by the State and transformed into conventional warfare, Ernst also explores the intricate relationship between the political essence of art and the idea of war concerning both urban life and the wilderness. From this perspective, Deleuze and Guattari's collaborative book projects can be viewed as Dadaist to a certain extent, reminiscent of the collective nature of Dada book projects, or the collaborative creations of Eluard and Ernst (Parish, 96).

Farewell My Beautiful Land of Marie Laurencin (1919) is one of Ernst's first war machines. Similar to other pieces from the same period, like *Portable Handbook* or *The Canalization of Refrigerated Gas*, Ernst endeavors to craft complex machines deliberately disorganized and dismantled, rendering any practical utilization impossible. In this sense, they embody true war machines; rather than functioning correctly to achieve their aim of destroying external objects, they self-destruct through their

[9] "Dada était une bombe" (Ernst 1970, 411).

malfunction. Let us refer to a complex non-utilizable machine that destructs itself incessantly as monstrous. These machine-monsters evolve throughout Ernst's work in many facets. In them, we find a conversation between the destructive nature of war, in which Ernst had a traumatic involvement, and the technological development due to these war-related self-destructive (anti-)goals. This poses a paradox for Ernst that leads him to his general view about humanity and nature.

As a prominent figure in Dadaism, Ernst discerns a fundamental difference between revolt and revolution: "revolt and revolution are not the same thing. But if you have this strong feeling of the need of revolt, need of freedom, and you are born into a period where there are so many events invite you to get revolted … it is absolutely natural that the work you could use is a revolutionary work".[10] By employing the concept of synthesis between subjective and objective lives, revolt represents the subjective counterpart that objectifies itself into revolution. Revolt emerges from within, often within a group, while revolution occurs as an event that is entirely unforeseeable and external. Revolt not only characterizes the spirit of twentieth-century French art but also embodies the subjective essence of French political thought—the very core of revolution. The so-called synthesis is, in fact, the fusion of the peculiar sensations within us and the absolute externality that abruptly occurs to us.

Ernst, as the artist of this synthesis, depicts the revolt of things, of birds, woods, and cities, capturing the objective revolt as the essence of social revolution. He presents this political concept of revolt in juxtaposition with the technical nature of World War I, merging elements of surrealist painting and collage. He is obsessed to the figure of the cage and of the bird, symbolizing technics and life, as even in forest are the birds in most of the cases in the cage, as if the woods play the role of the cage. This can be interpreted to suggest that the machinization of life, the deployment of war machines, commenced long before the advent of human culture, long before the Celsius twentieth century. This obsession is evident not only in the subjects he painted but also in his shift from the technique of frottage to grattage, thus bridging the creative power of nature with its destructive force. Intriguingly, his technical machines are also war machines. And

[10] https://www.youtube.com/watch?v=XGY4LrRV_GA.

Ernst's war machines exhibit a deep intertwining of both technical and biological elements. A striking illustration of this concept is *Elephant Celebes* (1921), which portrays an animal-machine, a true monstrous hybrid of an elephant and a war machine, as well as a celibate machine, a self-sexual non-oedipal desiring machine (referring to Duchamp's "auto-eroticism", and the notion of "auto-eroticism" in Deleuze's *Logic of Sense*: Deleuze 2018, pp. 202–203). It also evokes submarine warfare, suggested by the presence of two fishes in the sky and the ominous disturbance on the right side above. And we know that submarine warfare is "a quintessential example of a war-machine" in *A Thousand Plateaus* (Holland 2013, 43). This synthesis between technics and war, invention and destruction, reflects the political climate of Ernst's era: the destructive war in Germany and the constructive revolts in France. Ernst defines this synthesis not only as the core of art and a fundamental aspect of nature but also as the central focus of any exploration into the concept of freedom in the context of order and disorder. He reveals the inherent destructive forces within any constructive movement, and vice versa. In this context, the central question posed by Ernst's art is: what kind of cage is it, where there is a bird? What kind of machine is it, where there is life?

Ernst's fascination with collage signifies his ontological approach, bringing diverse domains into dialogue and connection. This encompasses not only the synthesis of subjective and objective lives but also the fusion of various realms within natural sciences, art, poetry, physics, molecular biology and bio-mechanics. However, in line with the essence of art and the nature of existence itself, these syntheses always manifest as something monstrous. Unlike philosophy, art doesn't require synthesizing the heterogeneous into a homogeneous sphere. Compared to conceptual philosophy, art and literature are closer to the essence of existence, unless philosophy embraces the monstrous nature of existence. Philosophy must acknowledge that any notion of peaceful synthesis is merely illusory. In this context, there is no peaceful union in Ernst's art; the war keeps going even in Europe after the rain (*L'Europe après la pluie*, 1940–42). Collage, in this regard, signifies the synthetic aspect of Ernst's work, emphasizing a recurring tendency in his creations and within the artistic tradition to which he belongs. This tendency involves pushing art to its limits, transcending conventional artistic boundaries, erasing categorical distinctions between art, philosophy, politics, and sciences. Ernst merges all these realms within the act of creation, encompassing the human, animal, and botanic domains. In essence, Ernst unites art with political ontology.

In this context, we can endeavor to grasp Ernst's monstrous machines. In *Nature's Gift of Gneiss Lava Iceland Moss* (1920), we witness the coexistence of bio-machines and techno-machines, not merely juxtaposed but instead converging and giving rise to a novel category of machines: bio-technical. Veins and arteries entwine with pulleys and strings, crafting a monstrous collage. Another instance illustrating this concept is the painting with the whimsical title *The Gramineous Bicycle Garnished with Bells the Dappled Fire Damps and the Echinoderms Bending the Spine to Look for Caresses* (*La Biciclette graminée garnie de grelots les grisous grivelés et les échinodermes courbants l'échine pour quêter des caresses*; 1923). Here, microscopic machinery such as sprockets or pinwheels infiltrates cellular structures like plant cells or bacteria. Ernst merges organic cells and machinic structures to fashion astonishing microbiomachines. They embody molecular machines, revealing the productive and creative industrialization occurring at the microscopic level of nature. Ernst, already cognizant of nature's inventive power, uncovers the molecular dimension of nature as its genuine machinality. On closer examination, it becomes apparent that these entities are not just technical machines but also war machines, underscoring the fact that every organism in nature, from bacteria to large animals, invents weapons as part of its natural history. Hence, at the core of his comprehension, Ernst establishes a connection between the concepts of natural history and war. He unveils the fundamental war at the heart of nature and links this war to the broad concept of technicity within nature. In his perspective, molecules are indeed war machines. Organisms' metabolic functions should be interpreted as defense mechanisms, propelling them to arm themselves through the invention of technical warfare.

This fascination reaches its culmination in his later works, titled *Microbes*, which are miniature landscapes that can easily fit in one's pocket. Are they also war machines? Or perhaps apparatuses of espionage? Evidently, in these works, he displays an interest in shifting between scales and domains, seamlessly transitioning from expansive landscapes to the microscopic world. Conversely, *Maximiliana* is the title of a collection of drawings that exemplify Ernst's exploration of large-scale megamachines, planetary machinery, and at the extreme, cosmic machines. He aptly embodies a concept present in Deleuze and Guattari's *A Thousand Plateaus*, where they assert that the earth itself can be seen as a "giant molecule" (ATP, 45). This implies a similar mechanism to what is evident in Ernst's transitions between flora, fauna, human, and mineral lives, as well as between machines and organisms. As an intriguing example, Ernst's depictions of flora are non-arborescent, or in Deleuze and Guattari's terminology, they are rhizomatic. Ernst never draws trees, he always paints the forest. He says, "when I make something, it almost always becomes a forest". In Ernst's perspective, the forest transcends being a mere collection of trees ("you can't see the forest for the trees"). In his work, the collective essence of the forest takes precedence over the individuality of a single tree. Furthermore, he portrays the forest as an intermediate realm, blurring the boundaries between the living and the dead. This concept extends to his depiction of the city as a forest, where buildings and streets diverge from arborescent forms (*Forêt et colombe*, 1927). The forest and the city, transcending the dichotomy between the living and the dead, are not organisms; forming striated spaces, they are machines, opposed to the sea as a smooth space (Holland 2013, 43). For both Ernst and Deleuze and Guattari, striation of a smooth space never ends up in an organism. This statement by Deleuze and Guattari encapsulates the essence of Ernst's work: "The wisdom of the plants: even when they have roots, there is always an outside where they form a rhizome with something else—with the wind, an animal, human being" (ATP, 12). This can be extended to the artistic deconstruction of everything in diverse fields: even when humans possess reason, art emerges to rescue them from dying of the truth (Nietzsche 1968, §822, p. 435).

Max Ernst's Anti-Oedipus

Automatism is a fundamental aspect of Dadaism, wherein the writing-machine endeavors to break free from the constraints of the human psyche, asserting its independence. By employing the technique of decalcomania, pioneered by Óscar Domínguez, Ernst propelled Dadaism into the domain of natural automatism, infusing it with the machinic essence of nature. The automation of writing no longer relies solely on collages of newspaper pieces; it now expands to encompass the arche-writing of nature—a liberating interplay of textures, colors, and forces.

However, with Ernst, we witness a dialectic between Dadaism and Surrealism. In Surrealism, this automatism or machinism becomes inherent in the psyche itself. Now, the psyche, or the brain, functions as an autonomous machine, as a plastic entity that both forms and adopts forms (Malabou 2008, 5). But how can art approach the full autonomy of psyche? As Warlick in *Max Ernst and Alchemy* observes, Ernst, since his youth, "was amazed by paintings and bread-dough sculptures created by the [psychiatric] patients. He found these works strangely alive and disturbing, and he intended to write a book about the art of the mentally ill" (Warlick, 35). Similarly, Turpin mentions that "As a student of psychology Ernst had visited a local mental hospital, where paintings produced by the inmates had caught his imagination" (Turpin, 4). But none of these commentators mention the schizophrenic origins of Ernst's art. In a manner similar to Deleuze and Guattari who build their *Anti-Oedipus*, among other things, on the artistic work of the psychotic Adolf Wölfli, Ernst finds psychiatry as an instrument to open up the unconscious mind to the exterior.

The idea of the exteriority of the unconscious is evident in *Oedipus Rex* from 1922. The nut, or the brain, is going to be opened outside, where the male and female birds are trapped. Ubl's formula that "nature is the subject's unconscious" (Ubl, 79), can be reversed, asserting that "the unconscious is nature's subjectivity". Therefore, the schizoanalysis of nature suggests a journey back from Surrealism to an absolutization of Dadaism, transitioning from the concept of the plastic brain to the idea of nature as a Giant Brain. Thus, Ernst applies the Dadaist machinism to all of nature, portraying it as a large writing-machine, a brain that generates dreams in an *Endless Night* (1940), the fantasies of nature as exemplified in *Jardin peuplé de chimères* (1936). The idea of the schizoanalysis of nature implies that the unconscious is fully externalized and materialized, while politically opposing the social structure of western culture. Ernst transforms Oedipus into a bronze sculpture, an animal figure, or even multiples of them—petrified Oedipus, signifying a metastable equilibrium between two archetypal figures: mommy and daddy. These dual aspects are embodied in *Oedipus I* and *Oedipus II* (1934), translating the machinic unconscious into the realm of the societal unconscious. Ernst's performance here, as expected, is ironic, indicating the petrification of Oedipus' journey at its very departure point. These dual figures of Oedipus are comparable with *Little Machine Constructed by Minimax Dadamax in Person* (1919–1920), a writing-machine (organs becoming letters) and oedipal machine (organs becoming genitals), revealing the synthetic and heterogeneous nature of both the machine and the unconscious. Ernst elevates the psychoanalytic dimension of surrealism to a non-human, natural, ontological plane, thereby de-oedipalizing the machinic unconscious. He relocates the psychoanalytic dimension of surrealism into the realm of nature, thus initiating an authentic schizoanalysis of nature. This suggests that the de-oedipalization of surrealism involves its transition toward the realm of molecular machinic nature. In this context, Ernst reconstructs the figures of Oedipus, diverting them from their Freudian psychoanalytic construction and, instead, aligning them with the archaic, primitive roots where social connections and lineages closely mirror natural molecular bonds. Therefore, symbolization in his work transcends the dichotomy of mind and nature, evolving into an embodiment of nature itself. In this context, objects symbolize each other in a material manner, where the bird's head symbolizes the sun and the cage symbolizes the jungle. This is the essence of how mythologies function. Through Ernst's work, surrealism retreats from Freudian psychoanalysis and reverts to its mythic origins, emancipating the Oedipus Complex from Freudianism and thereby laying the foundation for a schizoanalysis of nature.

References

Brassier, Ray. "Concrete Rules and Abstract Machines: Form and Function in *A Thousand Plateaus*". In: *A Thousand Plateaus and Philosophy*. Edited by Henry Somers-Hall, Jeffrey A. Bell and James Williams. Edinburgh University Press, 2018.

Breton, André. *The Lost Steps*. Trans. Mark Polizzotti. Lincoln: University of Nebraska Press, 1996.

Buchanan, Ian. *Deleuze and Guattari's Anti-Oedipus: A Reader's Guide*. Continuum, 2008.

Chakrabarty, Dipesh. "The Climate of History: Four Theses". *Critical Inquiry* 35 (Winter 2009).

Deleuze, Gilles. *Logic of Sense*. Translated by Constantin Boundas, Mark Lester and Charles Stivale. Bloomsbury Academic, 2018.

Deleuze, Gilles and Guattari, Félix. *Anti-Oedipus: Capitalism and Schizophrenia*. Trans. Robert Hurley, Mark Seem and Helen R. Lane. Minneapolis: University of Minnesota Press, 1983.

Deleuze, Gilles and Guattari, Félix. *A Thousand Plateaus: Capitalism and Schizophrenia*. Trans. Brian Massumi. London and New York: Continuum, 2004.

Deleuze, Gilles. *Francis Bacon: The Logic of Sensation*. Translated by Daniel W. Smith. Continuum, 2003.

Ernst, Max, *Beyond Painting: And Other Writings by the Artist and his Friends*, Wittenborn, Schultz, Inc. 1948.

Ernst, Max. *Ecritures*. Gallimard, 1970.

Holland, Eugene. *Deleuze and Guattari's Anti-Oedipus: Introduction to Schizoanalysis*. Routledge, 1999.

Holland, Eugene. *Deleuze and Guattari's A Thousand Plateaus*. Bloomsbury Academic, 2013.

Malabou, Catherine. *What Should We Do with Our Brain?* Translated by Sebastian Rand. Fordham University Press, 2008.

Malabou, Catherine. 2017. "The Brain of History, Or, the Mentality of the Anthropocene". *South Atlantic Quarterly* 1, 116: 39–53.

Nietzsche, Friedrich. *The Will to Power*. Translated by Walter Kaufmann and R. J. Hollingdale. New York: Vintage Books, 1968.

Parish, Nina. '"Pour faire un livre dadaïste": Dada Experimentation with Book Form', in *Dada and Beyond, Volume I: Dada Discourses*. Ed. Elza Adamowicz and Eric Robertson. Rodopi, 2011.

Turpin, Ian. *Max Ernst*. Phaidon Press Limited, 1979.

Warlick, M. E. *Max Ernst and Alchemy: A Magician in Search of Myth*. Austin: University of Texas Press, 2001.

Ubl, Ralph. *Prehistoric Future: Max Ernst and the Return of Painting between the Wars*. Translated by Elisabeth Tucker. Chicago and London: The University of Chicago Press, 2013.

Appendix B: Béla Tarr: Fascism and Cosmology

Cinema's *raison d'être* is beyond storytelling. At least after the war in Europe, cinema learned to think and to create cinematic thoughts. Narrative thinking has already been explored in the science of logic, its causal relations, and the laws of common sense. Cinema, in its youth, adhered to this logic of objects, subjects, actions and situations, marveling at the capabilities of montage, whether in the organicism of Griffith, which represents the common-sensical causal chains, or the dialecticism of Eisenstein, which represents the dynamic logic of concepts. Later, the reduction of cinema into an industry became intertwined with the standardization of montage, serving the conventional causal logic of "truthful narration", which unifies the characters and is directed towards a defined goal (*Cinema II*, 133). The cinema of long shot-close up-action is built upon the conventional teleo-logic of perception-recognition-decision (reaction). Nevertheless, genuine cinematic thinking does not necessarily follow this classic image of thought, causality and narrational temporality. Good films tell also stories, yet, as Godard famously noted, "not necessarily in that order", which means, not in the order of chronological time consisting of past, present and future, of recognizing and reacting, and of true narration. Great films strive to disrupt conventional logic,

M. Parsa, *Machinic Ontology*, https://doi.org/10.1007/978-3-031-99561-3

seeking ways to tell stories while avoiding a strict adherence to narrative and action, inventing temporal dimensions beyond narration and action.

Gilles Deleuze's *Cinema* books aim to present the post-war cinema as the emergence of pure cinematic thought and its non-chronological temporality. In *Cinema II*, Deleuze intriguingly asserts that "the essence of cinema—which is not the majority of films—has thought as its higher purpose, nothing but thought and its functioning" (*Cinema II*, 168). However, cinema approaches its essence through a political event. Shaped by the backdrop of World War II, Deleuze identifies a fundamental political dimension in his distinction between the pre-war classical cinema of action (movement-image) and the post-war cinema of thought (time-image). He defines this shift through the concept of the "unanimity of people". The post-war cinema is characterized by the dissolution of unified acting people and the rise of the divergence of thought. This transformation has not only impacted the themes of films but also influenced their structure and style, altering the roles of characters and reshaping the dynamic between films and their audiences. "A direct experience of time" is graspable in modern cinema through breaking from any totalization of images, people and the relation between man and the world (267). The emancipation of cinema from narrative time, and approaching the direct time-image, corresponds to liberating individuals from the totality of a people defined by concepts such as race, nation, social class, and so on. In this context, the cinematic revolution in narrative temporality, with the introduction of thought as the primary purpose of cinema, carries a political dimension that can be characterized in relation to manipulation, illusion, control, and in a word, fascism. As Evans and Reid assert, in Deleuze's perspective, post-war cinema can be viewed as an aesthetic endeavor aimed at freeing thought and its temporality from "the fascistic violence of imposing false unanimity upon peoples" (Evans and Reid 2013, 7). The role of cinema should not be to unite people and eliminate their unanimity through the imposition of a collective dream using narrative temporality. Instead, its purpose should be to stimulate thought in individuals in favor of an outside that dismantles their integrity. Dreams, imaginations, and identifications can lead viewers to surrender themselves to the logic of the film, making them susceptible to manipulation. On the contrary, thought instills resistance in viewers against mediocrity, common-sense, and the normalized dogmatic image. Modern cinema performs this resistance through distributing and expanding what David Rodowick in his *Gilles Deleuze's Time Machine* calls "irrational intervals" (Rodowick, 186). It is solely within these irrational intervals that one can engage in thought. The temporal logic of rational continuity and integration is what ought to be politically resisted through thought. In this sense, as Julian Reid observes, "Modern political cinema is, then according to Deleuze, a cinema of resistance to fascism, in which the idea of the people as a unified collectivity is itself an intolerable ideal" (Reid 2013, 78).

However, fascism emerges as a central theme, more so than in *Cinema* books, in Deleuze's collaboration with Félix Guattari in the two volumes of *Capitalism and Schizophrenia*. Within these volumes, fascism is examined not merely as a violent imposition of false unanimity on a people but as a psychoanalytical mechanism intricately woven into the functioning of collective unconscious desire. It signifies a socio-psychoanalytical mechanism whereby a community desires its own enslavement. The key feature of Deleuze and Guattari's exploration of fascism is to conceptualize it beyond the "antagonistic contradictions between the fascist Axis and the Allies", thereby transcending the schism imposed by WWII (Guattari 1995, 239–40). According to their perspective, fascism extends far beyond being merely a political ideology subject to historical analysis. They assert that neither solidarity nor historical identification can combat fascism, nor can they shield us from the insidious nature of molecular fascism.[1] Instead, Deleuze and Guattari seek to redefine the concept of desire and elucidate the fascist mechanism wherein collective desire turns against itself.

There is a limited body of scholarly works addressing a Deleuzoguattarian examination of cinema from a political perspective. In his introduction to *Deleuze and the Schizoanalysis of Cinema*, Ian Buchanan, in one of the

[1] Kriss Ravetto, in *The Unmaking of Fascist Aesthetics*, drawing from Deleuze and Guattari's insights, among others, discusses failed aesthetic endeavors in the post-war era aimed at "salvaging bourgeois moralism… and capitalist institutions from the stain of fascism" (Ravetto 6). Throughout her work, she adeptly illustrates how mainstream anti-fascist aesthetics, particularly in films, employ pre-fascist "narratives of evil" to represent fascism as a historically bounded and simultaneously an absolute entity.

rare studies on Deleuze and Guattari's volumes of *Capitalism and Schizophrenia* in relation to cinema, explores the potential of cinema to transcend its mediocre reality. The objective is to awaken this potential that stands in direct opposition to the *reality* of cinema as an instrument in the service of fascist manipulation and propaganda (Buchanan, 7–8).[2] The question is: What kind of films bring forth and actualize this potential of cinema in resisting the fascist tendency of collective desire? Deleuze himself, in *Cinema II*, introduces the cinema of time-image as emblematic of the political cinema of thought. In contrast, the collection of articles edited by Buchanan is replete with references to films that clearly align with what is termed by Deleuze the cinema of action-image or the cinema of thematization and recognition. For example, the exploration of the "irrational cut" as discussed in Gregg Lambert's chapter in this book, within the context of Stanley Kubrick's films, seems, in my opinion, to veer in a misguided direction as it confines itself to a symbolic realm: "... functions as ..." (in Buchanan, 29–30). "Irrational cut" should not be thematized or symbolized; rather, it should be executed and disseminated cinematically through the lens of the direct time-image. Pure cinematic thought cannot be thematized. To put it simply, the task of cinema is not to depict mad characters such as General Ripper but unveil the madness of pure seeing (32). Furthermore, at the political level, neither in Buchanan's introduction nor in its inner chapters does the book engage in a schizoanalysis of cinema in terms of *Anti-Oedipus*, the first volume of *Capitalism and Schizophrenia*, as a manual against fascism. This omission is particularly notable concerning the political essence of cinema, with thought as its primary purpose. The schizoanalysis of cinema should establish a connection between this essence and the concept of fascism as a mechanism of collective desire.

Similarly, when Deleuze defines modern political cinema by the absence of a unified people, it can imply that this cinema doesn't focus on

[2] Lorenzo Fabbri's "Neorealism as Ideology: Bazin, Deleuze, and the Avoidance of Fascism" is another study that examines Deleuze's perspective on cinema in relation to fascism. However, it specifically deals with Italian Neorealism and Italian Fascism, without addressing the broader concept of fascism as detailed in the volumes of *Capitalism and Schizophrenia*. In its exploration of Deleuze, the study concentrates solely on the *Cinema* volumes, overlooking the significant contributions of Deleuze and Guattari to the notion of fascism in the *Capitalism and Schizophrenia* volumes (Fabbri 2015).

describing an existing lifestyle of a people, but rather with individuals or communities that cannot integrate into a people. This makes some commentators to introduce the notion of "minor cinema", including films that depict minorities, as exemplifying Deleuze's modern political cinema. Once more, from my perspective, this seems to be heading in a misguided direction. For example, David Martin-Jones in *Deleuze Reframed* begins his discussion with Deleuze and Guattari's *Kafka: Toward a Minor Literature*, where the minor is not the matter of minorities but rather the matter of using the major language with a minor key, as is the case with Kafka's use of German language (Martin-Jones 2008, 54). However, he betrays his initial insight by dealing with the cinema of minorities in terms of discussing films that tell the story of minorities with using the apparatus of conventional cinema of movement-image, empty of pure political thought or direct time-image. His example of Gregg Araki's *Mysterious Skin* (2004) as a Deleuzian minor film is inappropriate because this film adheres to conventional storytelling methods (major key) while depicting the lives of marginalized individuals (queers) within an established genre: "New Queer Cinema" (58). In contrast to Deleuze's examples of modern cinema in *Cinema II*, which include figures like Vittorio De Sica or Alain Resnais, Araki's film predominantly serves the established identity of the queer community, particularly embodied in the character of Eric Preston played by Jeffrey Licon. Unlike the openness to the ambiguity of thought portrayed in films like *Umberto D* or *Hiroshima, Mon Amour*, Araki's work tends to reinforce existing identities rather than challenging them. Even thematically, the structure of *Mysterious Skin* remains within the framework of hiding and then revealing the truth about well-defined characters who happen to be queer or a-sexual. This truth-revealing logic serves to "heal" the minor status of the characters (64), aligning with the stereotypical Hollywood structure reminiscent of Hitchcock's *Marnie*. As David Rodowick aptly highlights, in reference to Syberberg's *Our Hitler* and Deleuze's analysis of it, "This is neither Hitler the individual, who is long gone, nor the totality that could put these images in a causal relation and so produce a coherent historical 'truth'. The power of Syberberg's film is a power of the false" (Rodowick, 187). The power of modern political cinema lies in its ability to present a complexity that defies totalization, resisting reduction to a narrative of lost and recycled truths. It embodies the pure power of the false, as appears in all of Godard's films. The pure form of time must intercede, creating an unhealable wound that permeates the subject and spreads throughout the external world.

In this context, my objective here is to position three of Béla Tarr's late films as prime examples that showcase political resistance against fascism through pure thought and its genuine temporality. Béla Tarr is the quintessential filmmaker of the stoppage of time and disruption of narration.[3] His films are far removed from being symbolic or thematic; they are irrational cuts extended to infinity. The deceleration of time and narrative in his later films not only represent an aesthetic style but also carry profound political significance. My claim here is that, Tarr's style of filmmaking attempts to liberate cinema from fascism in all of its forms, and this necessitates a modification in the field in which events and characters take place. While there are stories in his films, they are deliberately disintegrated, leaving behind personages and characters that lack an organic relationship with each other. Tarr crafted films devoid of action, scarcely punctuated by dialogue. Instead, they feature extended monologues or prolonged sequences of waiting, wandering, or peering through windows. "All stories are about disintegration". This quote from *Damnation* (1988) elucidates well Béla Tarr's cinematic world.[4] He employs cinema in opposition to integration concerning collective identities, or, in Deleuze's terms, the unanimity of a people (*Cinema II*, 216). His films tear apart both characters and viewers, dismantling the narrative coherence of society and the world. Disintegration emerges from the intrusion of the unnarratable and ungraspable reality into the closure of the story, lending a cosmological significance to every scene.[5] We attempt to

[3] Botz-Bornstein refers to the suspension of narrative time in Tarr's films as "organic time" (see Botz-Bornstein, 30). He derives the notion of organic, among other things, from the German Romantic tradition, exemplified by figures such as Schlegel and Goethe (4). However, as he acknowledges, this contradicts Deleuze's usage of the term in his Cinema books, where it primarily pertains to the movement-image rather than the time-image (5–6). Hence, we find the term "organic time" inappropriate in this context.

[4] Corinne Maury in her chapter on Béla Tarr relates the "disintegrating experience" in his films with how "waiting" of the characters functions as "breaking-down and self-denial" (Maury, 159).

[5] Compare with Collin Gardner's perspective on how Abbas Kiarostami draws inspiration from poems, rather than novels, as the foundation of his filmmaking approach, results in the dissolution

understand this account of reality in his films in terms of Deleuze's idea of thought as outside, only graspable through the direct time-image. Tarr abstains from telling stories because he believes that narrating through cuts and actions hinders the film's ability to approach real people and real situations. In a so-called Hollywood film, individuals and situations often serve merely as instruments for a rather mundane objective: entertainment. Through the suspension of narrative time, by prohibiting acting based on a script and instead urging actors to simply be and live the situations provided to them with some cards, and by minimizing the organic connections between characters, Tarr establishes a profound relationship between the audience and the characters in his films. The characters cease to be mere actors and instead transform into seers and thinkers. "This is a cinema of the seer and no longer of the agent" (2). In the signature scenes of his films after *Damnation*, when the character gazes out through the window into the empty exterior for an extended period, what truly unfolds is our journey into the character's inner universe. In this manner, Tarr gradually dissolves the distance between the role an actor plays and their real existence, aiming to approach the essence of being human isolated from social roles.

In this manner, Tarr creates films that are simultaneously political and cosmological. What renders political resistance possible, namely the disintegrated impersonal time, is identical to that which leads us to the absolute outside, namely the pure form of thought. In this essay, we aim to expound upon this connection by examining the role of the concept of fascism in three of his films, which are respectively *Sátántangó* (1994), *Werckmeister Harmonies* (2000), and *The Turin Horse* (2011). Drawing from Deleuze and Guattari's *Capitalism and Schizophrenia*, we characterize fascism as the inclination of people to be controlled and manipulated by a charismatic leader. For this, the leader requires a story, a story that

of narrative time and a move towards the direct time-image: "Kiarostami himself acknowledges this influence, arguing that, unlike the novel, the poem is inherently incomplete and non-linear, riddled with gaps and fissures that can defy convenient interpretation, not unlike Deleuze's 'irrational cuts' as temporal interstices in the time-image" (Gardner, 285). While Tarr's seminal films are based on Krasznahorkai's novels, Tarr's cinematic adaptation of these novels follows a similar trajectory to Kiarostami's poetic cinema.

encompasses a promise.[6] This takes place always within a problematic circumstance fraught with crises, disappointment and helplessness. The promise operates through belief, and the collective belief is intricately tied to the coherence and persistence of the story that fascinates the people. Within the story one believes, and the act of thought itself, in the Cartesian sense of doubting and counter-belief, assumes a political significance. We are controlled by stories; they are the obstacles that prevent us from thinking. In this context, any exertion of totalitarian power over individuals or a collective qualifies as a form of fascism if it manipulates not their lives but rather their desire. Tarr, by disintegrating the checkpoints of control-temporality governed by the rhythm of cut-action and introducing pure cinematic time grounded not in belief but in thought, signifies a profound resistance against molecular forms of fascism.

Therefore, in what follows, we employ the term fascism to denote a tendency that prompts us to cherish power, admire charismatic figures, and willingly submit ourselves to them. We will demonstrate that all these elements are evident in the films of Bela Tarr under consideration, especially in *Sátántangó* and *Werckmeister Harmonies*. Furthermore, we will see that, with Tarr, the concept of fascism transcends human relations and extends to relations in general and life itself. When life itself becomes tainted by fascism, resistance takes the radical form of cosmological extinction, a radical form of nihilism. In this regard, even the thought of the death of nature takes on a political and revolutionary sense. In my perspective, this is precisely what characterizes Tarr's cinematic journey, who initiates his career by scrutinizing human relations and subsequently transforms his work more and more cosmological without leaving his political departure point. This tendency culminates in his last and final film, *The Turin Horse*, that we will analyze in comparison with Lars von Trier's *Melancholia* from the same year.

[6] Deleuze in *Cinema II* clearly mentions the inherent relationship between storytelling and fascism, when he puts together "Hitler and Hollywood, Hollywood and Hitler" (*Cinema II*, 164). This assertion by Deleuze aligns with his collaborative effort with Guattari in examining the traces of fascism within post-war capitalism.

Sátántangó: The Cosmic Economy

Sátántangó is a cinematographic exploration into the nature of time in relation to swindle, belief and manipulation. Time, in *Sátántangó*, is acyclic and intricately interwoven with recording devices, memories, places and perspectives. This renders the temporal progression of the film not only weighty and slow but also irreducibly complex. The linear development, which is merely an outcome or a reduction of the real heterogenous temporality, takes the form of six steps forward and six steps backward, structured according to the chapters of Krasznahorkai's novel, from which the film is adapted. Linear time seeks to advance, yet the real temporality of characters and events contends against the overarching narrative. Hence, at least in the first six chapters, the viewer must consistently inquire: when and where does this take place? This doesn't refer to the specific time or location. Instead, it prompts considerations of whether it occurs before or after another presented perspective, the duration involved, the distance from another introduced viewpoint, and the time it takes to traverse that distance. Therefore, both time and space are relative and interdependent, in an Einsteinian sense. Each perspective derives its temporal and positional value from other viewpoints. Contrary to András Kovács' reading that adheres to chronological measurable temporality of *what time is it?* (Kovács 125–127), the key here is locating the relative position of each episode in relation with another. An event occurs, and its time is multiplied and distributed among the participants and their perspectives. This precisely constitutes the event itself; it is essentially defined by its relational temporality.

Irimias and Petrina are returning to the village from a nearby town after spending several years in prison, during which the villagers had believed them to be dead. A simple event. Everyone is connected to this event and, as a result, occupies a specific position in relation to it. This is how the event extends and multiplies itself. This positionality finds expression through emotions such as fear, anxiety, care, and other human sentiments. For some villagers, this constitutes a life event, something they will have to recount for the rest of their lives. However, only two characters appear to transcend this relational dynamic: the young girl,

Estike, and the doctor. One takes the objective form of belief, absolutizing the villagers' position in relation to Irimias, while the other adopts the subjective form of judgment, unraveling the entire story. Estike is the absolute object, believing literally everything, while the doctor is the absolute subject, doubting everything due to his infinitely poor memory. He relies solely on isolated facts, as the absence of a mind imbued with *logos* prevents these facts from coalescing into a coherent argument. Both are reduced to human machines, one functioning as an objective machine, and the other as a subjective machine. If the film were to have a focal point, it would be the encounter between Estike and the doctor on a rainy night beside the bar, the episode of the thought of suicide. Estike, who takes the life of her cat and then her own, stands as a counterpart to Edmund in *Germany Year Zero*, who ends his father's life before taking his own. However, what truly renders these two children as mirror images is not only their actions but their haunting journey towards death; Estike's childish beliefs mirror Edmund's playfulness in this journey. Edmund is convinced by the realist, or rather naturalist, arguments of his teacher to kill his father. Estike is fooled by his brother and by believing in angels and paradise. Should Estike be deemed an idiot and if she mirrors the character of Edmund, as suggested by Rancière, she does not embody idiocy in opposition to the rationality of others and their realist arguments (Rancière, 42). Rather, she represents the idiocy of reason.

The doctor, the man (or the God) who has factual knowledge about everything, is a stinky huge alcoholic. He is the one who measures time, embodying positivist science. The knowledge is exclusively in written form, comprised of scattered data interconnected through propositions of fact ("Futaki fears of death". "K behaved otherwise in last Autumn"). They are data that cannot be totalized into information, barely reaching the form of an argument, and thus do not serve to provide any truth about the characters and events. The memory is sufficient only for the time-distance between seeing and writing. It enables the creation of a drawing and the comparison of it with old drawings of the same object. Observing the same scene from different perspectives relativizes the factual nature of the event, transforming the fact into information. Time and the location of measurement mutually influence and modify each other, giving rise to a space-time that proliferates through events and characters. The time of the event is no longer punctual but is distributed among perceptual and recording devices, as well as spectators, or more precisely, observers. The fly on the camera insists that we are one of the spectators and there is no absolute perspective (Futaki observes through Mrs. Schmidt's window; Futaki is observed through the doctor's window; the doctor covers his window in the final scene of the film, as if we have been witnessing everything through his window, etc.). The long takes, and the length of the film, operate the same mechanism between the short-term memory of the spectator and the recorded long-term memory of the film. The personage does not receive his personality only in relation to other personages, as is the case in mainstream cinema. The long take, and the visual and acoustic proximity, help giving a corporeal and emotional presence of the personage as a living being. Therefore, one can make a distinction, in the universe of the film, between the relationality of characters and the absolutization of persons in relation to the exterior world. The latter refers to human dignity, which is insistently irrelevant of the relational character of each person, the money in their pocket, the job with which they introduce themselves, the social positioning, etc. Corresponding to the distinction between characters and personages, there is a distinction between two accounts of time. Let us dare to refer to these accounts in terms of *chronos* and *aion*. The first is the relatively measured space-time, multiplied among the observers, while the latter is an infinite straight line of unmeasurable time symbolized by Estike walking towards death with the dead cat in her left arm.[7]

The film makes a synthesis between *chronos* and *aion*, the relative and the absolute, by demonstrating how the relativity of the recording devices and observers can lead to a commonality which is *a priori* to them, how different perspectives evolve into a transcendental disjunctive synthesis. The cosmos, the philosophical alternative to God, must be represented by an alcoholic with short-term memory. In such a scenario, where God is

[7] Paula Marrati's description of Edmund's walk towards death in her reading of Rossellini's *Germany, Year Zero* is perfectly compatible with Estike's suicide scene: "we can draw no conclusions about his feelings, nor can we decide between indifference, pain or cruelty: this is the child's business and his secret… The child among ruins, people and things abandon him, and we see his 'pensive' face, without knowing what is on his mind. It is only after the fact, after Edmund throws himself from the top of the bombed-out building, that we finally understand" (Marrati, 56).

omniscient yet not omnipotent[8]—far from being the "principle or master of disjunctive syllogism" (LoS, 294), and humans are manipulable machines, order is inevitable. The existence of a spiritless God, a divine evil, is the necessary condition for the transition from chaos to cosmos. Only a structure-power, operating with the force of law, possesses the capacity to forcefully converge divergent perspectives.

The problem of commonality underscores the political significance of the film. It revolves around a communal farm owned by the villagers, which is sold, and the proceeds from the sale are what brought Irimias and Petrina back to the village. This orchestrates a game involving the psychological positioning of characters in relation to each other within a social-political field. *Sátántangó* is a film about the molecular origins of fascism. In this regard, each character in the film is defined not only by their temporal relationality, but also its power-positionality. This implies that society is organized through the power dynamics each character holds over others—be it the little Estike over the cat, the brother over the sister, Irimias over everyone, and local government over Irimias, each characterized by distinct modes and affections. The time-relation is linked to power-structure via the concept of law. Law is an instrument to complicate and manipulate affections. In this sense, any social law must also be a moral law. The Captain, who represents the law, says to Irimias und Petrina, "it [the decision] depends to what mood I have" and then argues that, "it is not a personal thing. It is the matter of law". The crux of the issue lies in the dialectic between personal power and the impersonality, neutrality, and brutality of the law, a dynamic that directly impacts individuals. The concept of power-structure operates solely through a process of abstraction that isolates decisions and emotions, beliefs and thoughts.

Let us review the argumentation of the Captain in length: he says, "keeping order may seem the responsibility of authorities. But lately, it's the business of the people, the order. In freedom there is something which

[8] Botz-Bornstein pays attention to this aspect in the figure of the doctor in the film: "The doctor is also the only 'authority' in the sense of an author. It is indeed possible to interpret the principle story of *Satantango* as written by the doctor. However, in reality, he has no authority at all but is losing more and more of it, as the cooperative system of which he was once part has been dismantled. Now he is weak and decadent, and he does not even have a name: what kind of 'author' is this?" (69).

is non-human, but rather divine. But our lives are too short to find out this properly. If you search for a link between order and freedom, think of Pericles. He said, order and freedom are linked by passion. We have to believe in both as we suffer from both. Both form order and freedom. But human life is meaningful, rich, beautiful and filthy. It encompasses everything. But we abuse freedom, wasting it as if it were just some junk. People don't like freedom. They are afraid of it. This is strange because there is nothing to fear when it comes to freedom. Order, on the other hand, can often be frightening. But I must draw your attention to certain points. The fact of the matter is, you have no choice but to cooperate. And my point is that, if you really respect the law, them, it is obvious, that I'm just a little ahead of you in this. Practically, you have offered it".

After this monologue of the Captain, Irimias says, "we do respect you a lot Captain". The captain responds, "Not 'captain', the law!". Irimias: "but you are the law itself, Captain". "Oh, no! the law that's all of us together", and maintains, "from now on, you either work for me, or … well, I have to say, you really have no choice". This emphasis on the identity of a triangle, comprising the ruler, the commonality, and the law, reveals the structure of abstraction inherent in the operation of power and law. In this operation, freedom is solely attributed to divinity and is not accessible to real human beings. In heavens it might be otherwise, but on the earth, God is stinky, memory is short-term, and order is inevitable. Man is the wolf to man, and he must be governed, whether democratically or by sovereign power. The only way to live together is to submit oneself to authorities, because "you either work for me, or … [the other choice is not mundane but rather divine and therefore is not a choice]". Similar to Pericles, democracy serves merely as a tool to persuade people to submit themselves.

In a quaint countryside, close to a town, a handful of individuals live a post-catastrophic life. They sell their primary means of agricultural production (the cows) and opt to disband the collective farm, choosing to depart from the village. One potential reason could be the impact of the industrial developments, which is exemplified in the film with the bureaucratic presence of the nearby town and the factory, on the basics of their once-agrarian lifestyle. What remains for them is their primitive sociality, no longer able to be structured by economic production. They desire to depart, yet they don't know where to go. They can only dream of a comfortable life where they can soak their feet in warm water while gazing at their beautiful farms. This is the dream. The reality is that the minimums of a social structure are collapsed in the countryside. There are betrayal and swindle among them. The villagers have the money of selling the collective farm but they don't know how that money can help them to realize their dreams. In this situation, Irimias, who was one of these people, returns from the town after years of absence, shocking the villagers who believed he was dead. This sparks both fear and excitement among them. In their eyes, we observe a blend of fear and hope: Irimias has the power to either jeopardize their dreams by taking the money or, alternatively, assist them in realizing their aspirations. The intentions in Irimias' mind remain uncertain to the villagers. First of all, he employs his charisma and moral-metaphysical arguments on these individuals who are willing to give all what they have to earn a bit more of hope. Irimias presents a very convincing argument beside the lifeless body of Estike: if you are innocent, then who is this little girl? But do the villagers really need this argument when they are already convinced by having no other choice, by their fear and hopelessness? It doesn't matter at all if they are convinced. It doesn't matter if they believe him. Even, it doesn't matter that Futaki doubts. By the way, Irimias continues to use arguments as rhetorical tools. The only difference is the sense of freedom in their minds, although they all know they have no other choice, as Irimias emphasizes, "from now on you are free". He is simply replicating what the Captain did to him. *You are free but you have no choice.* This is how structure-power functions: The Spider's Work! We are entrapped although we are free to go anywhere we want. The psychoanalytical barrier of Luis Buñuel's *The Exterminating Angel* takes on a political dimension here.

The common property must be ruled. Money brings freedom, but in itself it has no structuring power. The villagers surrender their money and freedom to Irimias, gaining in return a plan and structure. It doesn't matter whether Irimias really has a plan or if he is merely playing the role of a sovereign ruler without any actual plan. The people need, or rather, desire to be ruled when, as Erich Fromm explains, they are weak and bored.[9] They are perfectly prepared to be subjected to a sadistic/masochistic power-structure. They are free, but their freedom is rendered meaningless when it leads to nothing more than drinking and forgetting. They have money, but in this situation, what is the utility of money except for the capability to squander and inflict harm upon themselves? If the system is so oppressive that there is no space for genuine fantasies and dreams, as emphasized by Béla Tarr in his interviews, stating, "no matter it is communist or capitalist, the system kills the fantasies and power of the people", then everything is primed for fascism. This is the conundrum of freedom, encapsulating the inherent connection between the law-structure and fascism. The villagers consent to being enslaved by Irimias. They thank him for deceiving and enslaving them.

Deleuze and Guattari, in *Anti-Oedipus*, which is known by Michel Foucault as "the guide for a non-fascist life", ask a question which is the key to approach the core of the film: "why do people still tolerate being humiliated and enslaved, to such a point, indeed, that they *actually want* humiliation and slavery not only for others but for themselves?" (AO, 29). In *A Thousand Plateaus*, Deleuze and Guattari answer this question by elaborating on the concept of molecular or micro-fascism. They reformulate the question as the following: "Why does desire desire its own repression, how can it desire its own repression? The masses certainly do not passively submit to power; nor do they 'want' to be repressed, in a kind of masochistic hysteria; nor are they tricked by an ideological lure" (ATP, 236–237). The conundrum arises from the fact that unstructured, free life is dreadful and even fatal, compelling the masses to transform themselves. No one can structure their life on their own. This is the conundrum of Robinson on the desert island. However, the same applies to social life. A social life lacking meaning and purpose is prepared to embrace a charismatic dictator as its ruler. Political fascism arises from what Deleuze in *Cinema II* refers to as "ordinary fascism" (*Cinema II*, 165). Therefore, there is only a slight hesitation among the villagers to subject and enslave themselves, not only to the charismatic ruler but also

[9] Refer to Chapter 5 of Fromm's *Escape from Freedom*, where he discusses how the feeling of weakness in people, through the psychological mechanisms of sadism and masochism, paves the way for authoritarian regimes (Fromm, pp. 163–165). See also pages 233–241, particularly page 236, for an exploration of how this results in the way people willingly submit themselves to a Fascist regime.

at a more profound molecular or ordinary level: "fascism is inseparable from a proliferation of molecular focuses in interaction, which skip from point to point, *before* beginning to resonate together in the National Socialist State. Rural fascism and city or neighborhood fascism, youth fascism and war veteran's fascism, fascism of the Left and fascism of the Right, fascism of the couple, family, school, and office" (ATP, 236).

A rural fascist state begins to emerge, akin to Pasolini's "fascism at bay" (AO, 175). Irimias takes the villagers to the town and disperses them as part of his plan. We all know what follows next. The last remnants of life in the village are gone. The doctor recognizes it. But after a stroll, he discovers that the catastrophe is much more profound than the extinction of social life in a remote village: "I must be crazy. I've mistaken the bells of the sky for the sound of the bell for the dead". It's not solely about the collapse of social life. The doctor recognizes that it is a cosmic catastrophe: "One morning, at the end of October, not long before the first drops of the insufferably long autumn rains fell on the parched sodic ground on the western side of the yard, when the mud of the stinking bog makes the roads impassable cutting of the town, until the first frosts, Futaki was awakened by the sound of bells. There was a solitary chapel eight kilometers to the south-west at the old Hochmeiss field. But it had no bell, its tower having collapsed during the war". Futaki is awakened by the sound of bells. But this is not from the chapel whose tower is ruined during the war. The catastrophe is not human; it's cosmic. In a situation where, as Rancière observes, "politics is reduced to manipulation" (Rancière, 3), resistance assumes the form of thought. The war is in the heart of nature, an absolute war, a total war. In response to Lenin, who in "The Fourth Anniversary of the October Revolution", in October 1921, writes, "This first victory is *not yet the final victory*, and it was achieved by our October Revolution at the price of incredible difficulties and hardships, at the price of unprecedented suffering, accompanied by a series of serious failures and mistakes on our part... The essential thing is that the ice has been broken, the road is open, the way has been shown" (Lenin, 56–57), Béla Tarr announces another revolution, a cosmic revolution, whose October comes before the cosmic ice. The doctor discovers the cosmic economy in response to Lenin's "economic construction, the laying of economic foundations for the new, socialist edifice on the site of the demolished feudal edifice and the semi-demolished capitalist edifice". Hence, *Sátántangó* as a cinematic investigation on the nature of time entangles with an investigation on the historical origin of fascism. After hours of six steps forward and six steps backward, and asking what comes before and what after, the film turns back to the time before Futaki's waking up at the outset, before and beyond the economic collapse of the feudalist and the capitalist edifices, of the socialist Hungary, taking us to the deep history of the Marxist economic theory.

In the eleventh chapter of the film, titled "Just Trouble and Work", the officers, as part of their routine duties, rephrase Irimias' poetic descriptions of the personalities of the villagers into dry official descriptions. The prophet (Irimias) is not the one who elevates ordinary people to the level of divinity but rather the one who facilitates the transformation of divinity into the mundane. This Kafkaesque spirit dominates the film, and even the entirety of Béla Tarr's art: *law is more powerful than humanity*; although Tarr always insists that he is not a pessimist and has a little hope for the triumph of humanity over the brutality of social structures. His optimism led him to create films, while his pessimism led him to decide not to make films anymore after *The Turin Horse*, a film on the end of the world.

Werckmeister Harmonies: The Cosmic Order

An eclipse: "A dramatic turn of events. The air suddenly turns cold, the sky darkens. The dogs begin to howl, the rabbits hunch down, and the deer run in panic, stampeding in fright. Even the birds are confused and head to roost in this eerie, incomprehensible dusk. And then complete silence. Everything that lives is still".

The truth of extinction, more terrifying than a simple extinction of everything, is revealed in what follows: "no need to fear, it's not over. For across the Sun's glowing sphere slowly the moon swims away. And the Sun once again bursts forth. And to the Earth slowly, there comes light again, and warmth again floods the Earth. Deep emotion pierces everywhere. They have escaped the weight of darkness". What is more terrible than nihilism is Nietzsche's thought of eternal recurrence, the "thought of thoughts" (in Heidegger 1990, 23). All fades into nothingness, a complete darkness, leaving no trace for the emergence of a new sun and new emotions. A transmogrification completely devoid of any semblance, without any lingering echo of the past life.

What constitutes natural harmonies, or rather, the harmonies of nature? With what rhythm does nature unfurl its cycles? *Werckmeister Harmonies* is a film about order and disorder in both nature and social-political life. It explores the role of power as the means of order and its impact on natural play of order and disorder. It examines how power triggers disorder to make its order more and more abstract and empty, isolating order from life. It scrutinizes into what madness and fascism in this relation mean. Through the film, we can clearly observe that fascism is a mass movement, intertwining with the ways people disseminate beliefs and fears among themselves. Simultaneously, the film orchestrates a conversation between the social-psychological facet of fascism and the cosmological-philosophical contemplations surrounding the complexities of order and disorder.

In a small town in Hungary, likely during the communist era, a cold winter coincides with the shortage of coal. At the same time, a circus, featuring a massive whale and a prince, arrives in the town. The mentality is unrest and it appears as though a political riot is imminent. Some inhabitants plan what appears to be a form of sabotage. János, an innocent (fool!) young postman, becomes entangled in the plot without a clear understanding of it. Ultimately, the state army intervenes in the town, quelling the outrage, resulting in casualties and the arrest of those behind the plot, including János, who finds himself confined to a mental asylum.

The film also portrays the reflections of Eszter, a musician dedicated to preserving natural harmonies disrupted by the system of Andreas Werckmeister, leaving what we commonly refer to as music bereft of its essence: "This is not a technical, but a philosophical question… On what do we base our belief that this harmony, the core of every masterpiece, referring to its own irrevocability, actually exists or not? From this it follows that we should speak of not research into music, but a unique realization of non-music which for centuries has been covered up and a dreadful scandal which we should disclose. Hence the shameful situation that all the intervals in the masterpieces of many centuries are false. Which means that music, with its harmony and echo, its unsurpassable enchantment, is entirely based on a false foundation". This is not merely a technical inquiry but a philosophical one. It raises questions about the possibility of living an entirely misguided life. What if the supposed natural order of our lives was never truly natural? What if we are being manipulated unknowingly? Could it be that our affection and admiration for the music we cherish stem from the subtle manipulation of our emotions by the composers? If so, our riot and rebellion would be nothing more than instruments serving a hidden prince. Exploiting crises, or more accurately, exacerbating crises for the sake of profit, employing destruction to augment the order of power—this is the operational mode of fascism.

This unfolds amid a context where disorder ostensibly prevails in the social life of the town, propelling its inhabitants into a state of turmoil. It is minus 17 degree of Celsius and there is shortage of coal. In the midst of this situation, a circus arrives in the city, with the huge stinky whale and a concealed mysterious prince. János, the innocent postman, observes the unfolding events but remains unaffected by the (dis)order of things and the beliefs and fears it instills in people. He sees without recognizing.[10] When asked, "how are things in the cosmos?" he calmly responds, "everything is fine", only to hear a woman's voice in the background declaring, "the world has gone completely mad". Intriguingly, the woman comes to the foreground while János goes to background, implying his ultimate fate. People talk about the prince all the time, start to believe his magic and extraordinary power, but János is amazed by the huge whale; the ideology that functions with beliefs and fears versus the wonder of nature: "a giant whale has arrived. This mysterious creature from the sea has come from the far-off oceans. Most definitely you have to see it too… Just see what a gigantic animal the Lord can create! How mysterious is the Lord that he amuses himself with such strange creatures"? It is the nature itself which is the most extraordinary thing. At the core of the film lies the contrast between the logic of nature, where the interplay of order

[10] Refer to Paula Marrati's account of the distinction between seeing and recognizing in the context of Italian Neorealism and Deleuze's "cinema of the seer" (58–59).

and disorder is a magnificent dance, as illustrated by the example of the eclipse, and the logic of social power, where order and disorder are organized by needs, beliefs, mistrust, and military force. Or in Gabriel Tarde's formulation, this is a distinction between the flow of desire, exemplified in the dancing nature and the mesmerized János, and the flow of belief, exemplified in the outraging mob.[11]

This contrast is also evident in the distinction between modern musical harmonies, which forcefully impose equal distances to tonalities, and the natural harmonies of music predating Pythagoras and Aristoxenu. This signifies the advent of classical music in a broad sense, echoing the essence of modernity and its humanistic values. Eszter, whom János deeply respects, contends that we are dancing to the wrong rhythm and have forgotten the natural harmonies where differences are appreciated. Therefore, the prince can effortlessly manipulate the outrage, leading them to believe that the patients in the hospital are the root cause of all problems. The march of the mob embodies not only the idealized image of fascism where people achieve complete unanimity, as Deleuze observes in *Cinema II* (216), but also illustrates how the hidden Prince, the political Werckmeister, imposes equal distances on the tonalities of individuals, rendering them devoid of emotions. This transition from aesthetics to politics underscores the translation of the distinction between natural harmonies and fascist order. Yes, one of the messages of *Werckmeister Harmonies* is that classical music, in a certain sense, is fascist. And the same holds to any art or industry, including cinema, that coercively equalizes the distances between the tonalities of its audience. Can one interpret this distinction in terms of the deleuzoguattarian distinction between major and minor, the musical terms receiving political signification? Is the position of Jazz in the America by Afro-Americans similar to Kafka's use of German language in relation to Czech language?

In the state of absence of natural harmonies, military force becomes the sole embodiment of order. We find ourselves ensnared in the dualism of military order and natural chaos. This is because people fail to perceive that the Lord performs all His deeds solely for his own amusement.

---

[11] Tarde, Gabriel. *La logique social*. Paris, Alcan, 1893.

People are too serious in János' eyes; life is too hard. János himself embodies rather Dostoyevsky's idiot, inherently a visionary, "a seer, a perfect 'Idiot'" (*Cinema II*, 176). It comes as no surprise that Tarr chose to make the film only after encountering Lars Rudolph, whose gaze had the ability to embody the character of János as a pure seer of the intolerable. Whether labeled as innocent or idiotic, thinking courageously requires transcending the seriousness of one's relationship with life. Why can't we simply wait until the eclipse concludes, and the planets and the sun resume their innocent dance in harmony? Why can't we forget entirely, like children, and embrace a fresh beginning? Similar to Nietzsche, János deems it impossible to regard life in this world as a mere childish game meant for amusement, to live a simply amazing life. Similar to Nietzsche, he finds himself in a life devoid of vitality in an asylum, relinquishing both speech and existence, following a conviction for his alleged involvement in the plot. The film reveals the fate of a seer, among the community of agents, who refuses the power of action in order to remain a visionary but ends up being deprived from the power of sight (126). In contrast to conventional political films, the role of a visionary in this movie does not lead to the creation of a people or their liberation, as seen in films like *V for Vendetta* (2005), for instance. János ends up in a mental hospital similar to the one the mob attacks, beating the insane, and finally faces a naked old man who exemplifies the pure life—a life that has no place in the serious enterprises of the social order. The arrogance of man, aspiring to claim realms once believed to be exclusive to gods, contrasts with the pure life that combats this arrogance through ignorance. "God is the greatest of planners",[12] indicating that there is no hero, and the visionary is not a prophet.

Is the whale only for covering up all the villainy that "they" plan to do, or is the cause of all these miseries? As Jacques Rancière in *The Time After* observes, the whale is there to divide two orders, the social order of manipulators and manipulated and the cosmic order of János and Eszter who "lack the possibility of compromise" (Rancière, 54–55). The whale and the naked old man show two aspects of the same thing: Life on earth

[12] *Holy Quran*, Surah Al Anfal, 30.

is evil. The homeless, those who owe nothing, the ill, the ugly—they are the ones embodying pure humanity in a world where Tarr employed his camera as a tool, or perhaps a weapon, to alter it. Yet, he gradually realized that the pain is cosmic. But Tarr is not completely pessimist. The old man triumphs over the mob, and despite János ending up in the psychiatric asylum and the circus being ruined, the whale persists at the center of the marketplace, witnessed by Eszter as a reminder of cosmic harmonies. Or to be more precise, the eye of the whale. As János's gaze fades away in the end, the sorrowful look of the whale endures before Eszter, serving as the ultimate symbol of human dignity, or the dignity of life. Eszter's look, at the end of the films, along with the eye of the whale, materialize the last remainder of a little hope, making *Werckmeister Harmonies* the ideal "cinema of the seer". It is a film about the juncture of eye of the whale, the look of Eszter, and the extinguished gaze of János, implying that humanity attains its dignity at the juncture of its animality. Tarr's response to the fascist problematic does not involve a socialist state giving power to the humiliated people while maintaining the existing evaluative structure. Instead, he introduces a new ethics in which every single entity gains dignity through its very existence. If Tarr is a philosopher, he is undeniably a cynic philosopher.

The Turin Horse: The Cosmic Shit

The wood worms stop eating. The horse refuses to eat and drink. The well dries up. The light darkens. The wind quiets down. Nature ceases to fight. Everything in nature understands what is happening. Humans very late, after all of them. Nietzsche sees the horse and suddenly understands why it does not want to move. He hugs the horse: "*Mutter, ich bin dumm*". He then gives up speaking, writing and living, joining the rest of nature. Eleven years life without vitality; the penalty for not being so clever to call himself a fool at least once a month.

Béla Tarr, in an interview, says, "when you are young, you think that there are only social problems. Once we solve those, life will be a bliss and paradise comes. But later you realize that these are not the ontological problems. It's not as simple as that. And then a new phase comes, when you realize that the shit is cosmic".[13] The question is whether Tarr's films are philosophical or political, or perhaps he engages in philosophizing the political issues through the medium of cinema. After all, he never wanted to be a filmmaker; he wanted to become a philosopher. His later films, including the final one, *The Turin Horse*, depict an apocalyptic world where the political-economic crisis is entwined with a cosmic catastrophe. He places the small family outside the social-political context and situates them within the cosmological framework, within the fundamental and material conditions of life. Life is a struggle against death. When the battle ceases, there is neither life nor death. The cosmic shit is the omnipresent possibility of universal extinction, acknowledging the possibility that the wind might not have started blowing in the first place. The question is no longer, "what's this war in the heart of nature?"[14] The question is, "what if the war stops?" The question is not, what is the power-structure at play in a specific case. The question is, what if the forces have no power at all? It is not, how life can emerge amidst the dead nature. It is, what if nature dies.

If Nietzsche's philosophy can be summarized as exploring how power, stemming from the forces of life, can manipulate life itself, diminish its potentials, and ultimately killing the life itself, *The Turin Horse* would be an account of Nietzschean nihilism. Therefore, the transition from the young Tarr, interested in social-political issues, to the older Tarr, who approaches problems more cosmically, can be viewed in the context of a shift from the role of the State (in pseudo-communist Hungary) to the role of the market (in post-Soviet Europe) as that which targets not only the form of social life but rather the molecular fabric of life itself. It involves molecular fascism, spreading itself within the framework of ordinary market economy, as symbolized in the concluding conversation of Pasolini's *Salò*: "What is the name of your girlfriend? – Margherita". What does the logic of capitalism have to do with a cosmic ontology? Is it the same message, introduces by Fredrik Jameson and elaborated by Mark Fisher, that "it is easier to imagine the end of the world than the end of capitalism?" (Fisher, 2) Is it true that if someone rejects this

[13] https://www.youtube.com/watch?v=UhwHoCEYvMo.

[14] *The Thin Red Line* (1988).

system, their only recourse is to wish for total extinction as the sole hope? Is the cosmification of the shit linked to global capitalism?

Under capitalism, the social structure of power is amalgamated with, or perceived as a component of, the material-cosmic arrangement of forces. Béla Tarr, who, in the mentioned interview, states, "I still consider myself an anarchist", engages in a thought experiment in his later films: instead of the communist-utopian notion of a social-political revolution, he attempts to envision the death of nature and explores how this thought experiment can influence our conceptions of humanity and life. The motivation to radically transform social life, a defining aspect of communism, gives way to an anarchist pessimism: "this is the best of possible worlds, and it is a bad one". The political revolution transforms into the philosophical thought experiment of the death of nature. What is the meaning of life, considering the possibility of the death of nature?

The death of nature is the ultimate fate and indicates the universal meaning (or, meaninglessness) of everything. The difference between *The Turin Horse* and *The Werckmeister Harmonies* lies in the fact that in the former there is no day after the night, no bursting of the Sun after the eclipse, no eternal return of the same. Instead, Nietzsche's *Amor Fati*, the love of fate, materializes in *The Turin Horse* through the ungrounded decision of the father and the daughter to return back home, where the fire has no flame and the water is dried. Or rather, there is a ground; they return back home because "there is nowhere to go". The political barrier of *Sátántangó* (we are free to choose our enslavement) becomes fully cosmological here. The daughter says to the horse, "you have to eat". At the end of the film, the father says to her, "you have to eat", while trying to nibble an uncooked potato. She understands now the horse's decision. She refuses eating, and after her the father, the last being that understands what's going on, follows the way of the wood worm, the horse, and the daughter. The film that finalizes Tarr's filmmaking career, materializes successfully the seven days of extinction.

Lars von Trier's *Melancholia* is another film from the same year (2011), the year of the thought of extinction, in which nature understands the truth of extinction before humans, another film in which a horse refuses to move and is beaten, in which the small family is isolated from social and political issues and is situated within the basic conditions of life, life itself is reduced to a battle with death, a film in which movement to forward slows down and becomes impossible. Among the humans, only the melancholic Justine has a dark knowledge from nature within herself, in contrast to the scientific knowledge held by John, who remains ridiculously optimistic and exceedingly fragile when confronted with reality. Justine says: "The earth is evil. We don't need to grieve for it... Nobody will miss it... All I know is, Life on earth is evil". And she insists she knows that there is no other life anywhere else: "we are alone... Life exists only on earth. And not for long". She, who feels sick in the first part of the film amidst capitalist social life (due to having something from nature within herself), and gains strength and vitality in confronting nature's triumph over humans, says to her sister, "I'm very clean", after making love with Melancholia. But while Justine, in *Melancholia*, transforms the love of fate into a dramatic and erotic making love with the dancing death, embracing the total extinction, the unnamed girl in *The Turin Horse* becomes one of the particles of the dying nature, demonstrating an indifference which makes her position much more profound. *The Turin Horse* is a better film because it avoids the confusion of the thanatic event with erotic heroism. The gypsies say, "come with us to America!" The unnamed daughter does not want to be the cheerful Justine. Both of them are seers, but the significant difference lies in the fact that while Justin's seeing has an object, namely melancholia, the grown Estike in *The Turin Horse* sits before the window, observing an empty landscape. One can barely say that she sees. Justine evolves into a monotheistic believer in the unique God of extinction, while the unnamed girl, akin to Estike, believes literally anything, even the anti-bible of the gypsies. This elevates her position beyond that of Estike and Deleuze's "cinema of the seer", surpassing the context where even suicide becomes an action. Unlike Estike's act of consumption of poison or Edmund's act of throwing himself from a tall building, she ceases to live simply by abstaining from eating. This is not even a suicide.

This aligns with two senses of melancholia in relation to fascism: the melancholia of submission in von Trier, and the melancholia of resistance in Tarr. In the first case, Justine is repulsed by the market economy, exemplified by her employer and Michael at her wedding party. Summarizing in line with Fredrik Jameson and Mark Fisher, in order for imagining the end of capitalism, one has to imagine the end of the world. In the second case, if we can take *Werckmeister Harmonies*, which is based on Krasznahorkai's novel, *The Melancholy of Resistance*, as the prelude for *The Turin Horse*, the concern is not the market economy but rather a molecular fascism that permeates every individual and even every particle of the universe. Suppression no longer has an objective cause, it infuses itself within us and becomes depression. In this case, there is no planet to determine the fate of the world. The grown Estike, who killed herself once in *Sátántangó*, the unnamed girl, remains objectless, reaching the point of pure being.

In summary, we attempted to illustrate how Tarr's filmmaking style engages with the political core of thought, intertwining with an exploration of fascism in his films. His films perfectly align with Deleuze's conceptualization of cinema's political task, which revolves around thought as its primary purpose. Hence, we diverge from Andras Kovács' interpretation of Tarr's cinema, as expressed in his statement, "what the dramaturgical form of Tarr's films succeeds in doing is making the viewer believe the same illusion as the characters, which is that the story is going somewhere, even if slowly, while in fact it is only making a circle, and arrives at a situation as hopeless as the one it started out from … *the art of Bela Tarr is thus to make believe without hiding anything*" (Kovács 100; not my emphasis). We content that Tarr's work is not about belief but rather about thought. It is not concerned with illusion and the viewers' identification with characters (or "empathy", p. 130), but rather with thinking the situations and the times where the characters are involved. Likewise, my reading of Tarr's cinema opposes Clara Orban's interpretation, wherein Tarr's slow cinema "allows us to contemplate the places in his films" (Orban, 3). She, too, overlooks the profound political nature of thought in his films, instead emphasizing Tarr's "geographies of indifference" (12). Tarr's films are about thinking the times rather than contemplating the places.

Finally, Tarr elevates the political purpose of cinema to a cosmological realm. It is no longer about overcoming Fascist regimes; rather, it concerns a form of fascism that infiltrates the very fabric of life, extending beyond the confines of human social life. This aligns with what Evans and Reid refer to as "liberal fascism" (Evans and Reid 2013, 3). It is when the source of our slavery is not a concrete regime but is disseminated within the natural mechanisms that govern, not only our life, but that which is naturally subordinated. In liberalism, it is less about race or nationalism than is about those who are powerless, vulnerable, unattractive, alcoholic or those who may choose homelessness, perhaps as a means of avoiding submission to the power structure that would otherwise force them into being a docile member. It is their repression, their voices, their human dignity, and their political resistance which is targeted by Tarr, not only in his strong contribution in *Visions of Europe* (2004), but also in *Damnation* and *Sátántangó*. This also extends to the role of animals in his films. It is this melancholic resistance that drives János in *Werckmeister Harmonies* into madness. In a conversation alongside Rancière, Tarr asserts, "I don't like the word humanism because it is mostly used by people who are as least as possible humanist". Similarly, one might argue that the most profound manifestations of fascism should be sought not among those openly identifying as fascist, but rather among those who label themselves occasionally as anti-fascist while rigorously adhering to power structures.

References

Botz-Bornstein, Thorsten. *Organic Cinema: Film, Architecture and the Work of Bela Tarr*. Berghahn Books, 2017.

Buchanan, Ian and MacCormack, Patricia (ed.) *Deleuze and the Schizoanalysis of Cinema*. London and New York: Continuum, 2008.

Deleuze, Gilles and Guattari, Félix. *Anti-Oedipus: Capitalism and Schizophrenia*. Trans. Robert Hurley, Mark Seem and Helen R. Lane. Minneapolis: University of Minnesota Press, 1983.

Deleuze, Gilles and Guattari, Félix. *A Thousand Plateaus: Capitalism and Schizophrenia*. Trans. Brian Massumi. London and New York: Continuum, 2004.

Deleuze, Gilles. *Cinema II: The Time-Image*. Trans. Hugh Tomlinson and Robert Galeta Minneapolis: University of Minnesota Press, 1997.

Deleuze, Gilles. *The Logic of Sense*. Trans. Mark Lester. London: The Athlone Press, 1990.

Evans, Brad and Reid, Julian (eds.) *Deleuze and Fascism*. New York: Routledge, 2013.

Fabbri, Lorenzo. "Neorealism as Ideology: Bazin, Deleuze, and the Avoidance of Fascism". *The Italianist*, 35. 2, 182–201, June 2015.

Fisher, Mark. *Capitalist Realism: Is There No Alternative?* Zero Books. 2009.

Fromm, Erich. *Escape from Freedom*. New York: Discus/Avon Books, 1965.

Gardner, Colin. *Chaoid Cinema: Deleuze and Guattari and the Topological Vector of Silence*. Edinburgh University Press, 2021.

Guattari, Félix. *Chaosophy*. New York: Semiotext, 1995.

Kovács, András Bálint. *the cinema of BÉLA TARR the circle closes*. London and New York: Wallflower Press, 2013.

Heidegger, M., *Nietzsche. Vol. II: The Eternal Recurrence of the Same*, ed. D. F. Krell, New York: Harper San Francisco, 1990.

Maury, Corinne. *The Sense of Place in Contemporary Cinema*. Translated by Francis Guévremont. Edinburgh University Press, 2022.

Lenin, V. I., *Collected Works Volume 33*. London: Lawrence and Wishart, 1966.

Rancière, Jacques. *The Time After*. Minneapolis: Univocal Publishing. 2013.

Ravetto, Kriss. *The Unmaking of Fascist Aesthetics*. University of Minnesota Press, 2001.

Rodowick, D. N. *Gilles Deleuze's Time Machine*. Durham and London: Duke University Press, 1997.

Sutton, Damian and Martin-Jones, David. *Deleuze Reframed*. London and New York: I.B.Tauris & Co., 2008.

Orban, Clara. *Slow Places in Béla Tarr's Films: The Intersection of Geography, Ecology, and Slow Cinema*. London and New York: Lexington Books, 2021.

Tarde, Gabriel. *La logique social*. Paris, Alcan, 1893.

Author Index[1]

[1] Note: Page numbers followed by 'n' refer to notes.

M. Parsa, *Machinic Ontology*, https://doi.org/10.1007/978-3-031-99561-3

Subject Index[1]

[1] Note: Page numbers followed by 'n' refer to notes.

M. Parsa, *Machinic Ontology*, https://doi.org/10.1007/978-3-031-99561-3

I

J

L

M

N

The manufacturer's authorised representative in the EU is Springer Nature Customer Service Centre GmbH, Europaplatz 3, 69115 Heidelberg, Germany. If you have any concerns regarding our products, please contact ProductSafety@springernature.com

Printed and bound by CPI Group (UK) Ltd, Croydon, CR0 4YY
30/09/2026
02225583-0006